Some New Results

- *The development of a metatheory of physics theories based on the metatheory of mathematical logical-deductive systems.*

- *We develop a map between Gödel numbers and the set of elementary particle Lagrangians; and, in the event that the ultimate "Theory of Everything" is not describable by a lagrangian, we develop a map between the fundamental postulate(s) at the base of a non-lagrangian Theory of Everything.*

- *We prove that Gödel's Theorem implies that the Theory of Everything (the ultimate theory of Nature) must be a quantum theory. It cannot be deterministic.*

- *We show that the consistency of the Theory of Everything cannot be rigorously proven in principle. We further suggest that a Theory of Everything may be impossible in principle.*

- *We define quantum and classical probabilistic, non-deterministic Chomsky-like grammars. There appear to be two general types of probabilistic, non-deterministic grammars.*

- *We define **quantum Gödel numbers** for quantum Turing machines that embody quantum probability amplitudes.*

- *We define a map between Quantum Grammars and the elementary particle Standard Model. Particles form the alphabet; Lagrangian interaction terms define quantum probabilistic production rules.*

- *We define a map between Quantum Computers and Superstring Theories within the framework of a polycephalic (multi-tape) Quantum Computer formalism with the processor and memory having the character of fermions and tape heads having the character of bosons.*

- *We define a quantum computer formulation of quantum Assembly language and the quantum C language capable of extension to other computer languages such as C++, Java and Pascal.*

The Metatheory of Physics Theories, and the Theory of Everything as a Quantum Computer Language

Some Other Books By Stephen Blaha

Quantum Big Bang Cosmology: Complex Space-time General Relativity, Quantum Coordinates™ Dodecahedral Universe, Inflation, and New Spin 0, ½, 1 & 2 Tachyons & Imagyons™ (ISBN: 0974695815, Pingree-Hill Publishing, Auburn, NH, 2004)

Quantum Theory of the Third Kind: A New Type of Divergence-free Quantum Field Theory Supporting a Unified Standard Model of Elementary Particles and Quantum Gravity based on a New Method in the Calculus of Variations (ISBN: 0974695831, Pingree-Hill Publishing, Auburn, NH, 2005)

The Life Cycle of Civilizations (ISBN: 0972079580, Pingree-Hill Publishing, Auburn, NH, 2002)

Available on Amazon.com, bn.com and other Internet sites as well as at better bookstores (from Ingram distributors).

Cover Credits
Cover by Stephen Blaha © 2005.

The Metatheory of Physics Theories, and the Theory of Everything as a Quantum Computer Language

Stephen Blaha, Ph.D.[*]

Pingree–Hill Publishing

[*] sblaha777@yahoo.com

Publication Date: July, 2005

Pingree-Hill Publishing
P. O. Box 368
Auburn, NH 03032 USA

Or email to: baliltd@compuserve.com

ISBN: 0-9746958-4-X

This book is printed on acid free paper.

rev. 00/00/02

NOTE TO READERS

This book was written for a varied group of physicists, computer scientists, mathematicians and individuals in the general public. As a result parts of the book will seem too elementary to some. *The beginning chapters of the book will appear to be deceptively simple to the trained physicist. The concluding chapters of the book will be challenging for most readers.*

The best way to read the book is to skim the easy parts, to read the more challenging parts carefully but not to falter, and to enjoy the parts that suit your fancy.

The reader who is solely interested in the metatheory of physics theories can go immediately to chapter 13 since the material in the earlier chapters is not crucial to the discussion in chapter 13 and the succeeding chapters. The earlier chapters do establish a background in particle physics theories and quantum computers, and thus will increase the level of understanding of non-physicists who wish to better understand the physics described by the metatheory analysis.

PREFACE

This book is a substantially expanded version of *The Equivalence of Elementary Particle Theories and Computer Languages: Quantum Computers, Turing Machines, Standard Model, Superstring Theory, and a Proof that Gödel's Theorem Implies Nature Must Be Quantum.* Most sections have been revised and expanded and three new chapters have been added.

There are numerous physics theories for the many kinds of physical phenomena that exist in nature. Some theories are specific to a particular set of phenomena. An example is Newtonian mechanics, which describes the non-relativistic motion of objects. There is another set of theories, "Theories of Everything", that attempts to describe all phenomena in nature from a fundamental lagrangian or set of equations. Superstring theories and the new two-tier unified theories of the Standard Model and Quantum Gravity are examples of these universal theories. These theories are candidates for a "Theory of Everything." Weinberg (1992) has given a set of requirements for a Theory of Everything (which he calls a final theory): that it be quantum, contain symmetries, unify interactions, and so on. These requirements are sensible but they raise an important deeper issue: why these requirements?

As I. Rabi once said, "Who ordered this?" This question was also considered by Einstein when he raised the question in the form, Did God have a choice when he created the universe and its physical laws? Is there an arbitrariness in the choice of the Theory of Everything or is there only one reasonable choice for the Theory of Everything (which we will denote T_E hereafter).

The proponents of Superstring theories seem to feel that there is only one possible T_E (all others being ruled out by physical and internal consistency considerations) but the exact T_E eludes them. The aficionados of Standard Model variants and generalizations also have a similar feeling. The mood of theoretical particle physics suggests a tentative end is in sight and that a T_E is "just around the corner."

This feeling is reminiscent of physicists at the end of the 19th century when there were two forces of nature, electromagnetism and gravitation, electrons, and the periodic table of elements. The idea of unification of the forces in one theory was not an objective. The question of the existence of atoms was controversial since no one had seen one. And issues such as the spectra of the elements, blackbody radiation and so on were not viewed as critical. In view of this historical perspective it is this author's opinion that nature has more surprises in store for us and theorists today do not have sufficient information to construct a final T_E but may have sufficient information to construct interim approximations to a T_E.

Interestingly, there is a similar fundamental problem in Mathematics. Gödel's Undecidability Theorem shows that there is no one formalized system of logic that is adequate to handle all proofs in Mathematics. Thus a "Logic Theory of Everything" T_L is not known and one can crudely say that our understanding of logic and proof in Mathematics is incomplete.

Mathematicians have responded to their dilemma by creating a theory of mathematical proof that Hilbert called *metamathematics*, and that present-day mathematicians call the *Metatheory of Mathematics*. The metatheory of mathematics attempts to establish theorems about the nature and limits of mathematical proof. Gödel's Undecidability Theorem is perhaps its best-known result.

Given the current state of the search for the T_E it seems reasonable to develop a corresponding Metatheory of Physics with two goals: 1) to mathematically (and metamathematically) understand the nature and limitations of physics theories in general, and 2) to see if one can obtain general conditions that would be required of the

T_E (or, in other words, to answer Einstein's question – Are some or all of the features of the T_E totally fixed or are some (or all) of the features of the T_E arbitrary?). We will address the issues of a metatheory of physics after first considering current fundamental particle Theories of Everything. Then we will establish an equivalence between particle theories and quantum computer languages within the framework of quantum computers. We choose to follow this approach since it establishes quantum computers as a universal framework for the study of candidates for a T_E and sets the stage for the development of a metatheory of physics.

The basis for our view is the simple statement that computer data processing, and interactions between elementary particles, are the only phenomena in nature that represent pure creation and annihilation, and embody intrinsic transformations between input and output items. Computers can create bytes of data. Computers can transform bytes of data from one form to another. Elementary particle interactions can create new particles, can transform energy into matter (and vice versa), and can transform one type of particle into another. In all other situations where we see transformations – "what goes in comes out" – although the appearance of the output can be very different from the appearance of the input.

At a deeper level the bytes of information in a computer, and their transformations, have an associated thermodynamics that is similar to the thermodynamics of matter. And a deeper examination of elementary particle interactions shows that issues of information, and loss of information, are relevant in the quantum theories of their interactions.

Therefore we will examine candidate T_E's – the Standard Model and Superstring theories from the point of view of quantum Turing machines. Quantum Turing machines are formulated as in Blaha (2000). We will see how computer languages can be re-expressed in terms of quantum operators. A computer program then can be viewed as an evolving physical system. Superpositions of computer programs lead to superpositions of output quantum states. We will use Assembly language and the C language as examples.

To my wife Margaret

With Much Love

CONTENTS

1. Data and Particles

*The only acts of pure creation and annihilation occur
in computers and in elementary particle interactions.*

1.1 The Nature of Computer Data

Computers are so much a part of life today it is difficult to imagine that they embody deep concepts. We turn them on; play our computer games or read email; and don't think much of the activity transpiring within them. The development of the concepts of quantum computers has to some extent raised our level of interest – particularly among physical scientists.

The basic activity of computers is to take input "data particles" – bytes or bits of information, and create new "data-particles" of information – new data – through the execution of computer programs. A computer program implements the transformation of input data into output data.

A computer program corresponds to an evolution operator in particle physics. The evolution operator determines the change from the input particles in a particle interaction to the various possible sets of output particles.

What are these "particles" of computer data? Physically they are patterns of bits that are imprinted on magnetic media or optical media – merely patterns. Each bit is a number etched in the media having the value one or zero. A data value such as a number or a character of an alphabet is represented by a pattern of bits. It takes energy to create or change these patterns. So there is a thermodynamics of data just as there is a thermodynamics of matter.

The difference between data and matter is one of perception: we perceive matter directly with our senses of sight, touch and smell; we perceive data indirectly through electronic or optical means. This difference in perception is quite real in one sense and illusory in another sense. The reality of the difference is conveyed by a famous anecdote about Dr. Johnson, the great writer of the English Dictionary. Bishop Berkeley, an eminent philosopher of the eighteenth century, proposed the world that we experience with our senses is illusory and that reality actually consisted of Ideals – a set of forms or patterns that interact with each other according to the laws of nature. Thus he viewed the apparent solidity of material objects as an artifact of our senses. Dr. Johnson, upon being presented with this theory, said, "I refute it thus!" and proceeded to kick a rock.

Since data is embodied in patterns on some media it is conceptually similar to Berkeley's Ideals. And, as stated, data has features reminiscent of matter such as a thermodynamic theory of its transformations. Data can be created, transformed and destroyed just as elementary particles can be created, transformed and destroyed.

1.2 The Nature of Elementary Particles

Having established that data is really a set of patterns with some properties similar to matter, we now look at the question of what is matter?

Until the beginning of the twentieth century matter was thought to be indestructible. The law of the conservation of matter stated matter cannot be created or destroyed. Some scientists thought matter was continuous; some scientists thought matter consisted of atoms.

In the early twentieth century it became clear that matter consisted of atoms *and* that under certain conditions it might be possible to transform matter into energy or energy into matter. As the century progressed matter was then found to have two levels of substructure: protons, neutrons and so on which, in turn, were found to be composite particles composed of quarks.

Thus matter, and energy as well, was reduced to a set of fundamental particles that currently are around fifty in number: quarks, electrons, muons, gluons, photons, neutrinos, and so on. These particles however were somewhat strange. They were not "hard" little spheres like marbles. Rather they were a form or pattern of mass/energy that had both particle-like and wave-like properties. Sometimes they behaved like particles (typically when large numbers were involved); sometimes they behaved like waves (typically at high energies or short distances where quantum effects are significant). Some called these particles: wave-particles or names such as "wavicles."

If we examine the concept of these wave-particles whether in the Standard Model of Elementary Particles or in Superstring theory (where they are little mathematical strings or combinations thereof) we find that they do not have substance in the everyday sense of the substantiality of matter. Thus they are analogous to Berkeley's concept of Ideals.

Matter has dissolved at the sub-atomic level into insubstantial waves with particulate behavior under certain circumstances. What is waving? Don't ask. The waves are probability waves – whatever that means. There is a structure to particles. Particles embody data such as momentum and spin, and internal symmetry values – isospin, hypercharge and so on. Thus we can think of sub-atomic particles as data packets or data structures – patterns.

1.3 The Similarity of Data to Particles

At this point we have reduced matter to data packets at the subatomic level and established that computer data consists of sets of patterns with some properties similar to matter.

Both subatomic particles and data can be created or destroyed or combined to produce new particles or new data. The similarity at the present level of discussion is compelling. In the following chapters we will amplify these ideas at a deeper level. We will investigate particle interactions at a deeper level and show how to create "subatomic particle quantum Turing machines" for both the Standard Model and Superstring theories. (In this book we take a neutral stance on whether the Theory of Everything is a Superstring theory or some kind of quantum field theory.)

We will develop a theory of classical probabilistic Turing machines and quantum probabilistic Turing machines. We will show how particle interactions can be viewed as steps in a "computation." We will show how a quantum computer program in some computer language can be viewed as the evolution of a quantum physical system.

After establishing particle theories within a quantum computer framework, we will proceed to develop a theory of physics theories called a metatheory of physics. The metatheory of mathematics (metamathematics) is based on two main parts: a mathematical-deductive system such as elementary number theory, and a formal symbolic system, or calculus, that abstracts the logic of the mathematical-deductive system. Each theorem in the mathematical-deductive system corresponds to a theorem in the calculus. The calculus is in a sense the logical skeleton of the mathematical-deductive system.

The metatheory of physics that we will develop adds another system. The metatheory of physics consists of a mathematical-deductive system (a physics theory), a calculus – the logical skeleton of the physical theory, *and* the set of all possible statements (theorems) that are experimentally measurable either directly or indirectly. The consideration of the interrelations of these three systems will lead to a series of statements (metatheorems) about both physics theories limited to a domain of physical phenomena and the Theory of Everything. These statements will be about the nature of physics theories and thus independent of the current level of experimental knowledge. They will, in part, answer Einstein's question, "Did God have a choice in the design of the physical laws of the universe?"

Alonf the way we will see that we can associate Gödel numbers – numbers that embody mathematical proofs or programs – with particle theories. We will find that Gödel numbers can be associated with Quantum Turing machines and with Lagrangians in the set of possible elementary particle Lagrangians. (Lagrangians describe the interactions of particles.) And we will shows Gödel's Undecidability Theorem of 1931 implies the Theory of Everything must be quantum. Besides eliminating numerous attempts to create a deterministic theory underlying quantum theory we will have a proof of a major feature of physics.

2.The Standard Model of Elementary Particles

An elementary particle physics experiment is
equivalent to the execution of a Quantum Turing Machine.

Our understanding of the fundamental nature of matter today is in one way much more developed than that of 1900. In another way we still are at a stage where we think of particles of matter as fuzzy little balls. The balls are different – quarks, leptons, and so on in year 2005 instead of atoms of chemical elements in year 1900. And we know more about their properties. But they are still perceived as fuzzy little balls.

The accepted theory of the day is called the Standard Model[1]. It is an amalgam of earlier theories of electromagnetism, the weak interaction and the strong interaction. The Standard Model is consistent with almost all known experimental results. There are a number of variations of the Standard Model but they are all largely the same in overall characteristics. The combination of particles and interactions that constitutes the Standard Model is somewhat strange – there is no convincing explanation of the concatenation of features that it contains – yet it is also simple in many ways. Many of its features appear to be arbitrary – not logically justifiable. And it is known to be incomplete because it does not include gravitation.

The Standard Model started with the unification of the theory of Quantum Electrodynamics (electromagnetism) with the theory of the Weak interactions in a theory that is called the Electroweak Theory. This theory was developed by Shelley Glashow, Steven Weinberg, Abdus Salam and others. Afterwards the Electroweak theory was combined with the theory of the Strong interactions to produce the Standard Model.

2.1 The Families of Matter

Today the fundamental particles of nature, as described by the Standard Model, are grouped into three families: fermions, gauge field particles and Higgs field particles. Fermions are particles of half-integer spin that constitute what we ordinarily call matter. The fermions include electrons, quarks, muons, neutrinos and so on.

---

[1] There are many books on the Standard Model. See the References section.

Gauge field particles are particles that "carry" the interactions or forces between particles. You could call them "force field" particles because they embody the forces between particles. The photon is a gauge field. There are other gauge fields for the weak interaction and for the strong interaction.

Higgs field particles are heavy spin zero particles that are needed in the current formulation of the theory. They have not as yet been observed. The reason is thought to be their large mass. Particle accelerators are only now becoming capable of producing them if they exist. There are numerous speculative theories (such as Extended Technicolor theories) in which Higgs particles are not fundamental but are a form of composite particle. Since they have not been observed experimentally at the time of this writing there is no way of knowing.

The fermion family has two subfamilies: the quarks and the leptons. The distinguishing feature of quarks vs. leptons is that quarks can experience the strong force but leptons cannot. The lepton family includes electrons, muons (μ) and neutrinos (v). The members of the fermion family are listed in the following table:

The Fermion Family

Generation	Flavor	Quarks			Leptons	
I	1	up	u_1 u_2 u_3		v_e	electron neutrino
	2	down	d_1 d_2 d_3		e	electron
II	3	charmed	c_1 c_2 c_3		v_μ	muon neutrino
	4	strange	s_1 s_2 s_3		μ	muon
III	5	top	t_1 t_2 t_3		v_τ	tau neutrino
	6	bottom	b_1 b_2 b_3		τ	tau

Notice also the fermions appear in three similar generations or sets labeled I, II, and III. There is no obvious reason for this repetition of generations. We do not know at present why Nature has three generations rather than one generation (although some theories are better behaved if there are three generations). It is like having three sets of china when one set would do. Each generation contains 6 quarks and two leptons.

Each quark comes in three colors. The colors are labeled with the subscripts 1, 2 and 3. For example there are three "down" quarks – a triplet – that we have denoted d_1, d_2, and d_3 in the above table. These triplets are called *color triplets*. Quarks form triplets because the strong interaction has a symmetry called SU(3) that requires it.

Originally the three varieties of each quark type were labeled with colors such as red, white and blue. (The colors are merely symbolic. A "red" quark was not actually red. The name only served to distinguish between the quark varieties.) Now we use simple numeric subscripts. In fact these labels reflect internal quantum numbers that are called color quantum numbers by physicists. Consequently, the theory of the Strong

Interactions of the quarks is often called Quantum Chromodynamics since it involves the color quantum numbers.

The sets of color triplets are distinguished from each other by a quantum number called the flavor quantum number. Each triplet of quarks, and a corresponding lepton, has the same flavor quantum number.

The classification of quarks and leptons in the preceding fermion table is the result of over fifty years of experimental and theoretical analysis. It is like the periodic table of the elements that is so important for chemistry.

The twelve gauge fields ("force fields") in the Standard Model also have a pattern or classification scheme. This scheme is based on a form of mathematics called group theory. We will not go into the details of group theory as applied to elementary particles except to note that group theory symmetry is usually the form of symmetry that physicists refer to when they speak of the symmetries of a physical theory. Unlike the symmetries of a flower the symmetries of physical theories are not easy for non-physicists to appreciate.

The Gauge Field Table shows the gauge groups using a standard mathematical physics notation. The groups are all unitary groups (hence the "U" in each designation).

The Gauge Fields ("Force Fields") of the Standard Model

Gauge Group	Number and Symbols of Gauge "Particles"	
U(1) Weak Hypercharge	**1**	W_0 Name: W boson
SU(2) Weak Isospin	**3**	W_i $i = 1,2,3$ Names: W bosons
SU(3) Color (Strong)	**8**	G_i $i = 1,2, ..., 8$ Names: gluons

The weak hypercharge group is related to the strangeness quantum number discussed earlier. The weak isospin group and the weak hypercharge group are symmetries of the Electroweak theory. The color symmetry group is the symmetry of the strong interaction that binds quarks together to make up hadrons – particles such as protons, neutrons, pions, and so on. All the gauge particles in the table have spin one and are called bosons. There are two types of gauge bosons in the Standard Model: W bosons and gluon bosons. The photon is a combination of two W bosons.

The Standard Model incorporates all of our current experimental knowledge of particles and interactions (except gravity) in a single theory. The theory has been extremely successful in accounting for many experimental results although many interesting physical calculations in this theory cannot be done with current computational techniques (such as the binding of light quarks to produce protons, neutrons, and so on).

Many physicists feel the Standard Model agrees too well with the available experimental data. The lack of discrepancies and disagreements with experiment is a negative in a sense. Disagreements between experiment and theory are usually the source of advances in our understanding of nature. Grappling with discrepancies leads to new theoretical ideas. Like the ideal marriage with no arguments the success of the Standard Model without major experimental discrepancies makes life less interesting for the physicist.

2.2 Interactions in the Standard Model

The Standard Model is defined by a complex mathematical expression called a *Lagrangian* together with a procedure for quantization, and a procedure for the calculation of physically interesting quantities. The quantization and calculational procedures are specified by Quantum Field Theory. The details of these procedures are important but they are not relevant to the theme of this book.

The Lagrangian for the Standard Model can be divided into two parts. One part describes the behavior of non-interacting particles. These particles are called *free particles*.

The second part of the Lagrangian describes the interactions of particles. We can think of free particles as moving through space and interacting with other free particles through the exchange of gauge particles. Gauge particles are the particles that are the carriers of the electromagnetic, weak, and strong forces. The photon is the gauge particle for Quantum Electrodynamics (electromagnetism). Three W bosons are the gauge particles for the Weak Interaction. (The photon and the W particles are intertwined in the Electroweak theory.) Eight gluons are the gauge particles for Quantum Chromodynamics (the Strong Interaction).

The interaction part of the Standard Model specifics many interactions between particles. We will look at a representative sample of these interactions to obtain an understanding of their basic idea and then use these interactions to illustrate the linguistic representation of particle interactions. The linguistic approach will be explored in more detail in succeeding chapters.

The first interaction term in the Lagrangian that we will explore can be written in the form:

$$\bar{e}Ae$$

where e represents an electron (the bar over the left e signifies an "outgoing" electron in our simplified view) and A represents a photon. This term corresponds to a number of different interaction situations. The situations are depicted graphically in Fig. 2.2.1.

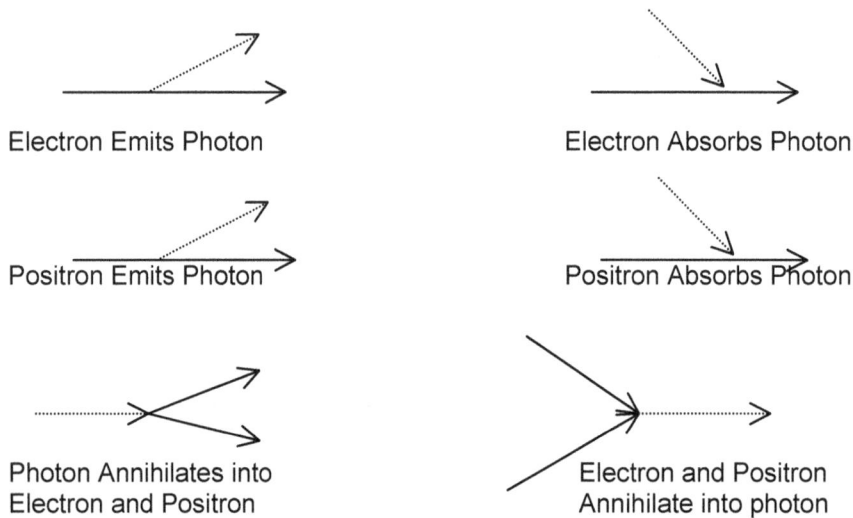

Electron Emits Photon

Electron Absorbs Photon

Positron Emits Photon

Positron Absorbs Photon

Photon Annihilates into
Electron and Positron

Electron and Positron
Annihilate into photon

Figure 2.2.1 Simplest Electromagnetic Interactions of an electron. Solid lines are electrons and positrons (anti-electrons). Dashed lines are photons.

The interactions depicted in Fig. 2.2.1 correspond to an amazing feature of matter: particle creation and annihilation with the conversion of energy into matter and vice versa. A photon (a particle of electromagnetic energy such as light) can transform into an electron and a positron creating matter from energy. An electron and positron can annihilate to form a photon thus creating energy from matter. Energy-matter conversion is theoretically based on both quantum theory and the theory of Special Relativity. Special Relativity is required because of the conversion between mass and energy ($E = mc^2$ at work). The creation and annihilation process is an inherently quantum process requiring Quantum Field theory.

The interaction processes in Fig. 2.2.1 are the simplest forms of the electromagnetic interactions of an electron (and positron). These simple interactions can be combined to make an infinity of more complex composite interactions. For example two electrons can scatter off each other by exchanging a photon. The photon exchange takes place by having one electron emit a photon and the other electron absorb it. This composite process can be pictured with a (Feynman) diagram such as:

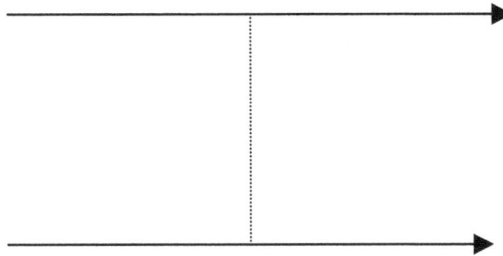

Figure 2.2.2 Feynman diagram for two electrons interacting by exchanging a photon (dotted line).

Another example resulting from the combination of the simple electromagnetic interactions is the case of two incoming electrons scattering (colliding) and producing three electrons and one positron (the output particles).

Electrons can interact by exchanging photons. Fig. 2.2.3 depicts how a pair of generated photons can combine to generate an outgoing electron and positron.

The preceding discussion shows how the simple basic interactions in the Standard Model Lagrangian can produce a complex variety of what we have called "composite interactions."

The diagrams that we have seen are used extensively in particle physics. They are called Feynman diagrams after their originator Richard Feynman. These diagrams correspond to very complex rules for performing calculations in quantum field theory. The rules form an approximation scheme for calculating the probability that a given process will take place. This approximation approach is called *perturbation theory*. Composite interactions are particular cases of perturbation theory.

For example, the calculation corresponding to the diagram in Fig. 2.2.3 together with calculations for other relevant diagrams, produces the probability that two electrons of specified energies and momenta will produce three electrons and a positron with specified energies and momenta. An experiment could be performed whereby electrons with the specified energies and momenta repeatedly collide with measurements made of the percentage of the outcomes that produce the specified output particles. The measured percentage, and the calculated probability multiplied by 100, should be the same.

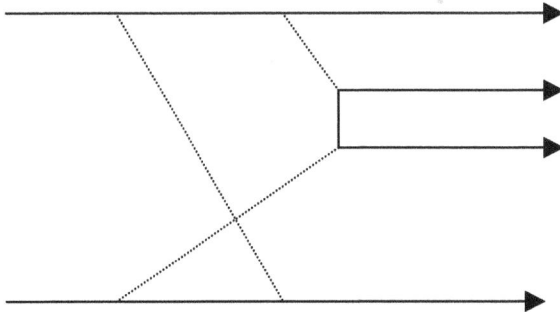

Figure 2.2.3 Two electrons collide producing 3 electrons and a positron. The dotted lines are photons.

2.3 Perturbation Theory and a Language for the Standard Model

In the preceding section we have considered electromagnetic interactions of electrons and positrons. The Standard Model contains many additional particles and also the weak and strong interactions. This brief view of some of the features of the Standard Model is meant to give the flavor of the features of the Standard Model so that a new linguistic view of the Standard Model can be presented. *In this new view the interaction terms in the Lagrangian specify the grammar of a computer language.*

Physicists use perturbation theory to perform computations in the Standard Model. Perturbation theory uses the interaction terms of the Lagrangian for approximate calculations of the probabilities of particle interactions. As we have seen perturbation theory calculations are normally visualized using Feynman diagrams. For example the collision of two electrons to produce two electrons with different energies and momenta (called electron-electron elastic scattering) can be visualized as a sum of terms corresponding to the various ways the electrons can interact. The number of terms is infinite in principle. Since this infinite sum cannot be calculated the sum is approximated by a finite number of terms. The simplest and, as it turns out, the dominant terms in electron-electron scattering are:

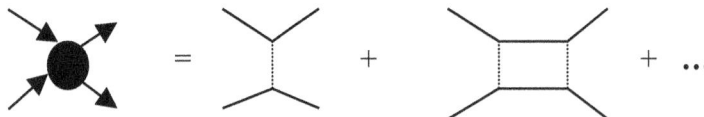

Figure 2.3.1 The first few terms in the approximate perturbation theory calculation of the scattering of two electrons. The dotted lines represent photons that carry the electromagnetic force between the electrons.

We will see that the individual terms in the perturbation theory calculation can be viewed as words. The words are part of a language with an alphabet and grammar.

The 36+ particles of the Standard Model constitute the set of symbols or the alphabet of the language. (It is an interesting, but meaningless, coincidence that most alphabet-based human languages have 20 to 40 letters in their alphabets. English has 26 and so on.) As we shall see, the grammar is a quantum extension of a type of computer language (developed by Chomsky and others) that uses *production rules*. The production rules for the grammar of the Standard Model are easily derived from the interaction terms of the Standard Model.

The concept of the language of the Standard Model is very simple. The technical details of the language description require a discussion of computer languages and Quantum Turing Machines. The next few chapters describe the basic idea of the linguistic representation of the Standard Model. They show how particle interactions can be viewed as transformations (processes) within a Quantum Computer that accepts the computer language generated by the Standard Model Lagrangian interaction terms.

3. A Linguistic Representation of the Standard Model

In the previous chapter we explored the features of the Standard Model. We saw that it was consistent with almost all the known properties of elementary particles. This chapter introduces a new view or representation of the Standard model that focuses on its interactions and *shows the Standard Model defines a language similar to a computer language.*

The alphabet of the language is the set of elementary particles of the Standard Model. The words of the language are quantum states consisting of elementary particles. A bound state of several particles such as a proton (three quarks bound together) is a word. A quantum state consisting of several free particles – particles that are not bound together and that are some distance from each other also constitute a word. In fact the entire universe constitutes one mighty word.

We shall see that the collision or scattering of particles can be viewed as beginning with a combination of letters corresponding to the set of initial particles – the input string. This input string undergoes transformations specified by grammar rules to produce an output string of letters corresponding to the outgoing particles after the collision.

In describing this new representation of the Standard Model we will focus on the essentials of the processes of creation, transformation and annihilation of matter ignoring (for the moment) particle spin, momentum, angular momentum and other details that are important in the complete theory of the Standard Model. This approximation may have been valid prior to the Big Bang when the universe might have been a mathematical point. (We will consider these "details" when we consider the SuperString Quantum Computer™. Incorporating particle spin, momentum, angular momentum and so on into a "particle" language is not difficult.)

The idea of associating physics with computers is not as unconventional as it might appear at first. Feynman[2] viewed computers as relevant for Physics: "If we suppose we know all the physical laws perfectly, of course we don't have to pay any attention to computers. It's interesting anyway to entertain oneself with the idea that we've got something to learn about physical laws; and if I take a relaxed view here ... I'll admit that we don't understand everything." Feynman wanted to simulate physics

---

[2] R. P. Feynman, International Journal of Theoretical Physics, **21**, 467 (1982).

computations on a quantum computer in the hope that it would be faster than a conventional computer. We will show the Standard Model itself actually defines a specific (theoretical) quantum computer – a far more exciting possibility – because it gives a new view of Reality. Nature itself is a form of computer.

SuperString theory can also be formulated within a Quantum Computer framework.

A computer language representation of particle physics is of great interest in itself. It may generate new insights into the process of matter creation and transformation. It may lead to a new understanding of the fundamental nature of the universe. And it appears to suggest a rationale for approaches such as the currently popular SuperString theories of elementary particles.

3.1 Linguistic View of an Interaction

We will begin by looking at the simple interaction term we explored in the previous chapter:

$$\overline{e}Ae$$

From a computer language perspective this Lagrangian interaction term can be viewed as specifying a set of grammar rules called *production rules*.

In fact each interaction term in the Standard Model Lagrangian can be viewed as specifying a set of grammar rules. The combined set of grammar rules for all Standard Model interaction terms defines a grammar with particles constituting the alphabet (letters or symbols) of the grammar.

To appreciate the mapping (or analogy) between particles and alphabetic letters, and of interaction terms and computer grammar, we have to understand the process of data characters (or letters) flowing through a computer. It is an interesting and little noted fact (because it is viewed as trivial) that a computer can generate (or absorb) data as part of the computation process. For example we might write a computer program that takes a set of letters input into a computer and outputs each input letter twice:

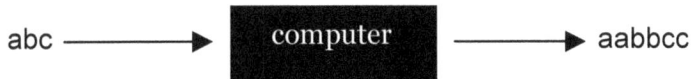

abc ⟶ | computer | ⟶ aabbcc

In a sense the computer has created data characters just like particle interactions can create particles. Computers can also absorb (or annihilate) data (usually to our dismay). So we can see an analogy between the transformations of data characters in a computer, and particle annihilation and creation. *Nothing else in Nature is so directly analogous to particle creation and annihilation.*

This observation leads us to take the view that particles are data packets that we denote with letters (symbols). They contain quantum numbers and other properties (mass, spin, momentum, and so on) that certainly are data. They have a grammar that we summarize with a Lagrangian.

3.2 Computer Grammars

The Standard Model Lagrangian in our view specifies a *grammar* in the sense of Naom Chomsky. Chomsky's concept of a language, and of a grammar, has important applications in the theory of computation and computers.

There are four basic types of languages in the Chomsky approach: called type 0, type 1, type 2 and type 3. They differ in the allowed forms of their grammar rules (also called *production rules*). We will be interested in type 0 languages. A type 0 language (also called an *unrestricted rewriting system*) is the most general type of language. It allows any grammar production rule of the form

$$x \longrightarrow y$$

where x and y are strings of characters.

Production rules specify how one string of characters transforms into another string of characters. Calculations in computers using computer languages are reducible to sets of grammar rules for string manipulation that are similar to the one shown above.

Each term in the interaction part of the Standard Model Lagrangian is equivalent to one or more production rules where the characters are particles. *The Standard Model can be viewed as generating a type 0 language.* This language goes beyond current types of grammars because it is inherently quantum probabilistic in nature. Quantum aspects of these rules will be described later.

Before looking at the production rules generated by an interaction term in a Lagrangian we will discuss a formal grammar. A grammar is a quadruple of items that is usually symbolized by the expression

$$<N, \ T, \ S, \ P>$$

where N is a set of variables called *nonterminal symbols*, T is a set of *terminal symbols*, S is a special nonterminal symbol called the *head* or *start symbol*, and P is a finite set of production rules. The angle braces < and > are merely a mathematician's way of saying these items are grouped together to constitute (or make) a grammar.

The terminal symbols are the set of characters that are allowed in input strings or output strings. The nonterminal symbols are the set of characters that appear in intermediate steps that lead from the input string to the output string. They are like internal variables or symbols. The combined set of terminal and nonterminal symbols make up the *vocabulary* (alphabet) of a language.

Chomsky's definition of a language is the set of all strings of terminal symbols that can be generated by applying the production rules to the head symbol (or start symbol) S. The head symbol is the symbol that begins all strings of symbols that can be generated in a language.

A simple example of a language in this approach is a vocabulary or alphabet consisting of the ABC's with words created from these letters according to some set of production rules.

3.3 Generalized Input Chomsky Languages

We will generalize Chomsky's idea of language to be the set of all strings that can be generated from all finite input strings of terminal symbols as well as the *head symbol*. We can also view all particles as generated directly or indirectly at the beginning on the universe. The "Big Bang" (the beginning of the universe) then becomes the primeval head symbol.

We can visualize the application of production rules to transform an input string of terminal characters into an output string of terminal characters as:

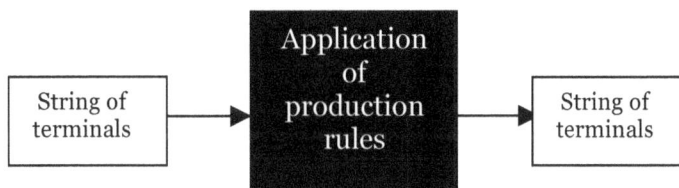

Figure 3.3.1 Generating an output string of terminal characters from an input string of terminal characters using the production rules of a grammar. Inside the black box, transformations of the input string take place and non-terminal symbols may appear and disappear. Non-terminal symbols, by definition, cannot appear in the input or output strings of characters.

In order to make these grammar concepts more concrete we will look at a simple artificial grammar before looking at the grammar generated by interaction terms in the Standard Model. The nonterminal symbols will be the letters S (the head symbol), A and B. The terminal symbols will be the letters x and y. The production rules will be

$$S \rightarrow AB \quad \text{Rule I}$$
$$A \rightarrow y \quad \text{Rule II}$$
$$A \rightarrow Ay \quad \text{Rule III}$$
$$B \rightarrow x \quad \text{Rule IV}$$

15

$$B \rightarrow Bx \quad \text{Rule V}$$

The Chomsky computer language that this grammar generates consists of all strings containing any number of y's followed by any number of x's since any of these strings can be generated from the head symbol S using the production rules. Because of rule I y's are placed to the left and x's are placed to the right. The order of the symbols matters just as it does in human language – consider the words ides and dies which differ only in the order of i and d!

An example of generating a string yyxxx from the head symbol is:

$$
\begin{array}{lll}
\text{S} & \rightarrow \text{AB} & \text{Rule I} \\
\text{AB} & \rightarrow \text{AyB} & \text{Rule III} \\
\text{AyB} & \rightarrow \text{yyB} & \text{Rule II} \\
\text{yyB} & \rightarrow \text{yyBx} & \text{Rule V} \\
\text{yyBx} & \rightarrow \text{yyBxx} & \text{Rule V} \\
\text{yyBxx} & \rightarrow \text{yyxxx} & \text{Rule IV}
\end{array}
$$

The production rule used to make each transition is listed above on the right.

Our generalization of the Chomsky definition of language would allow any string to be the starting point – not just the head symbol S. Using the sample grammar described on the previous page the generalized language becomes any string of x's and y's.

A more interesting language can be created by adding two new rules to the rules on the preceding page:

$$
\begin{array}{ll}
\text{y} \rightarrow \text{A} & \text{Rule VI} \\
\text{x} \rightarrow \text{B} & \text{Rule VII}
\end{array}
$$

The resulting language – the set of strings of terminal symbols – remains the same despite the addition of these new grammar rules. However the number and variety of transitions becomes much larger. For example the following chain of transitions is allowed,

$$\text{yx} \rightarrow \text{AB} \rightarrow \text{AyBx} \rightarrow \text{AyyBx} \rightarrow \text{AyyyBxx} \rightarrow \text{yyyyxx}$$

In the next chapter we will extend the concept of computer grammars by allowing probabilistic grammar rules – production rules which have an associated probability of executing.

4. Probabilistic Computer Grammars

Grammar, which knows how to control even kings.
Molière - Les Femmes Savantes (1672) Act II, Scene 6

4.1 Probabilistic Computer Grammars™

The preceding chapter described the production rules for a *deterministic grammar*. The left side of each production rule has one, and only one, possible transition.

Non-deterministic grammars allow two or more grammar rules to have the same left side, and different right sides. For example,

$$A \to y$$
$$A \to x$$

could both appear in a non-deterministic grammar.

Non-deterministic grammars can be easily (almost "naturally") associated with probabilities. The probabilities can be classical probabilities or quantum probabilities. An example of a simple non-deterministic grammar is specified by the production rules:

$$S \to xy \qquad \text{Rule I}$$
$$x \to xx \qquad \text{Rule II} \quad \text{Relative Probability} = .75$$
$$x \to xy \qquad \text{Rule III} \quad \text{Relative Probability} = .25$$
$$y \to yy \qquad \text{Rule IV}$$

where the head symbol is the letter S, and the terminal symbols are the letters x and y. The relative probability of generating the string xxy vs. the relative probability of generating the string xyy from the string xy is

$$xy \rightarrow xxy \quad \text{relative probability} = .75$$

$$vs.$$

$$xy \rightarrow xyy \quad \text{relative probability} = .25$$

The string xxy is three times more likely to be produced than the string xyy.

For each starting string one can obtain the relative probabilities that various possible output strings will be produced.

A more practical example of a Probabilistic Grammar™ can be abstracted from flipping coins – heads or tails occur with equal probability – 50-50. From this observation we can create a little Probabilistic Grammar™ for the case of flipping two coins. Let h represent heads and t represent tails. Then consider the grammar:

$$S \rightarrow hh$$
$$S \rightarrow tt$$
$$S \rightarrow ht$$
$$S \rightarrow th$$

$$h \rightarrow t \quad \text{probability} = .5 \ (50\%)$$
$$h \rightarrow h \quad \text{probability} = .5 \ (50\%)$$
$$t \rightarrow h \quad \text{probability} = .5 \ (50\%)$$
$$t \rightarrow t \quad \text{probability} = .5 \ (50\%)$$

The last four rules above embody the statement that flipping a coin yields heads or tails with equal probability (50% or .5).

Now let us consider starting with two heads hh. The possible outcomes and their probabilities are:

$$hh \rightarrow hh \quad \text{probability} = .5 * .5 = .25$$
$$hh \rightarrow th \quad \text{probability} = .5 * .5 = .25$$
$$hh \rightarrow ht \quad \text{probability} = .5 * .5 = .25$$
$$hh \rightarrow tt \quad \text{probability} = .5 * .5 = .25$$

If we don't care about the order of the output heads and tails, then the probability of flipping two heads and getting a head and tail (hh → ht or hh → th) is .25 + .25 = .5.

This simple example shows the basic thought process of a non-deterministic grammar with associated probabilities.

The combination of a non-deterministic grammar and an associated set of probabilities for transitions can be called a *Probabilistic Grammar™*. We will see that the grammar production rules for the Standard Model must be viewed as constituting a Probabilistic Grammar™ with one difference. The "square roots" of probabilities – probability amplitudes – are specified for the transitions in the grammar. The Standard Model requires probability amplitudes since it is a quantum theory. Therefore we will describe probabilistic grammars with associated probability amplitudes (such as that of the Standard Model) as *Quantum Probabilistic Grammars™*.

4.2 Quantum Probabilistic Grammar™

An example of a Quantum Probabilistic Grammar™ can be constructed based on an analogy with a two slit photon experiment. Imagine a wall with two slits. A source of photons shoots photons at the wall. A photon can go though either slit with equal quantum probability. An illustration of this experimental arrangement is:

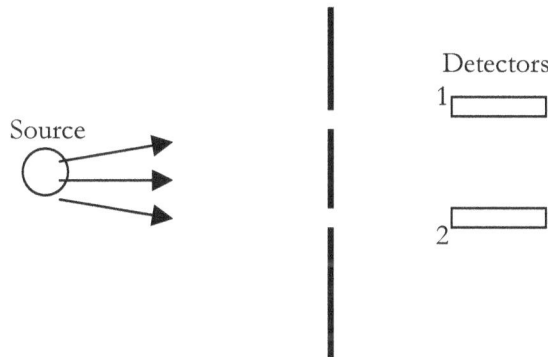

Figure 4.2.1 Two slit photon experimental setup.

A simple Quantum Probabilistic Grammar™ can be constructed corresponding to this experimental setup:

$$S \rightarrow 1 \text{ probability amplitude} = 1/\sqrt{2}$$
$$S \rightarrow 2 \text{ probability amplitude} = 1/\sqrt{2}$$

The head symbol S represents the source. The digit 1 represents a photon going through slit 1. The digit 2 represents a photon going through slit 2.

The values of the probability amplitudes $1/\sqrt{2}$ can be calculated using Quantum Mechanics. The probability for a photon to go through slit 1 is the absolute value squared of the probability amplitude:

$$\text{Probability to go through slit 1} = (1/\sqrt{2})^2 = .5$$

and the probability for a photon to go through slit 2 is

$$\text{Probability to go through slit 2} = (1/\sqrt{2})^2 = .5$$

This simple example illustrates the basics of a Quantum Probabilistic Grammar™.

Before applying these concepts to the Standard Model we will look at a simpler Quantum Field Theory called a ϕ^3 ("phi cubed") theory (ϕ is the Greek letter phi). This theory describes a self-interacting spin 0 particle with no internal symmetries. This theory is a stepping stone to the far more complex Standard Model Quantum Field Theory. We are only interested in it as a simple example of quantum probabilistic grammar rules.

The ϕ^3 theory is so named because it has a cubic Lagrangian interaction term. (Note the exponent 3.) The grammar rules for the ϕ^3 theory are:

$$\phi \rightarrow \phi\phi \quad \text{Rule I}$$
$$\phi\phi \rightarrow \phi \quad \text{Rule II}$$

Rule I corresponds to the emission of a ϕ particle and rule II corresponds to the absorption of a ϕ particle. We will not introduce a start symbol. Instead we will consider the transitions from an input state of a number of ϕ particles to an output state of (possibly) a different number of ϕ particles. *We will ignore the momenta of the particles*. (This assumption is equivalent to assuming the ϕ particles have infinite mass.)

We will assume either transition above takes place with a "relative probability amplitude" g. We will call this simplified theory the *modified ϕ^3 theory*. We will view g as a measure of the probability amplitude for an absorption or emission of a ϕ. (g is similar to a coupling constant in Quantum Field Theory.) The actual summed probability has to be normalized or rescaled so that the sum of probabilities equals one.

To get a feel for the Quantum Probabilistic Grammar™ approach we will look at the case of an input state consisting of two ϕ particles. The output states can have one ϕ, two ϕ's, three ϕ's, and so on. Each possible output state has a certain probability of occurring. The sum of the probabilities of producing all output states must equal one. (Remember that the sum of all possible outcomes of flipping a coin is one. Having it come up heads has probability ½ and having it come up as a tails has probability ½ also.)

The simplest string transition from a two ϕ "input" state to a one ϕ "output" state is:

$$\phi\phi \rightarrow \phi$$

using Rule II. The probability amplitude of this transition is g by assumption.

The transitions between strings can be visualized with diagrams that are like the Feynman diagrams that used in Quantum Field Theory perturbation theory calculations. These diagrams are not the same as Feynman diagrams because they embody time orderings of emissions and absorptions of ϕ particles. (They actually hark back to time-ordered diagrams used by physicists prior to 1950.)

In some simple cases the time ordering is irrelevant. For example, the Feynman-like diagram for the simplest case of a two ϕ input state transitioning *directly* to a one ϕ output state is the same as the Feynman diagram:

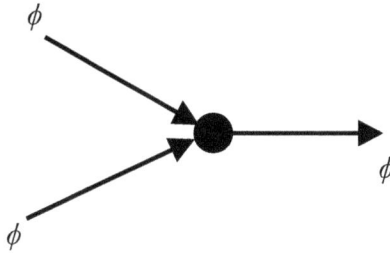

Figure 4.2.2 Diagram for $\phi\phi \rightarrow \phi$. The input states are always on the left and the output states are always on the right in Feynman and Feynman-like diagrams.

The time order of emission and absorption of ϕ particles can be symbolized using parentheses. For example,

$$(\phi)\phi \rightarrow (\phi\phi)\phi = \phi(\phi\phi) \rightarrow \phi(\phi) = \phi\phi \qquad \text{Diagram A (Fig. 4.2.3)}$$

and

$$\phi(\phi) \rightarrow \phi(\phi\phi) = (\phi\phi)\phi \rightarrow (\phi)\phi = \phi\phi \qquad \text{Diagram B (Fig. 4.2.3)}$$

These string transitions correspond to different time-ordered Feynman-like diagrams:

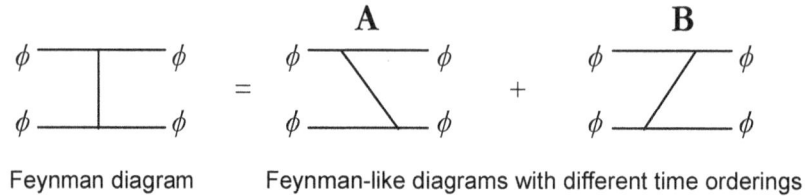

Feynman diagram   Feynman-like diagrams with different time orderings

Figure 4.2.3 The true Feynman diagram is the "sum" of the two time-ordered Feynman-like diagrams.

The correspondence between Feynman-like diagrams and the transitions between strings based on the Quantum Probabilistic Grammar™ can be seen by taking vertical slices on diagrams A or B above after each emission or absorption. For example,

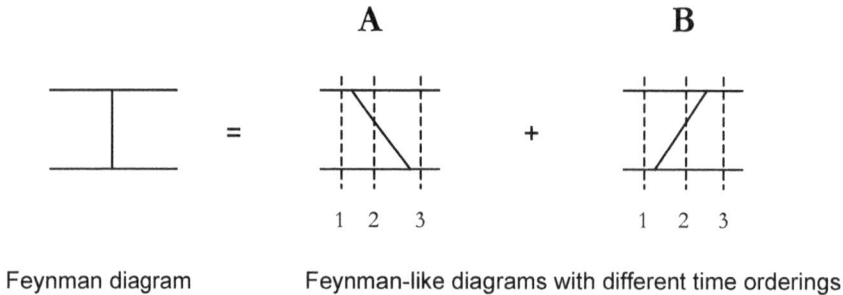

Feynman diagram   Feynman-like diagrams with different time orderings

Figure 4.2.4 Slicing Feynman-like diagrams. A slice is made after each emission or absorption. As you read down a slice the particles are listed in the same order as the corresponding string. Each slice is numbered starting from the left.

The string corresponding to each numbered slice in the above figures is similarly numbered in the following transitions:

Slice: 1   2   3
$$(\phi)\phi \;\rightarrow\; (\phi\phi)\phi = \phi(\phi\phi) \;\rightarrow\; \phi(\phi) = \phi\phi \qquad \textbf{A}$$

and

Slice: 1   2   3
$$\phi(\phi) \;\rightarrow\; \phi(\phi\phi) = (\phi\phi)\phi \;\rightarrow\; (\phi)\phi = \phi\phi \qquad \textbf{B}$$

Parentheses on the left side of an arrow indicate the particle(s) that emits a new particle(s) appearing within the corresponding parentheses on the right side of the arrow.

A transition from an input state containing ϕ particles to an output state containing ϕ particles always has an infinite number of ways of taking place and thus an infinite number of Feynman-like diagrams. Readers familiar with the perturbation theory of Quantum Field Theory will remember that these diagrams are the same as the Feynman diagrams generated by perturbation theory with the additional feature of having time orderings.

We will now look at the transition of two ϕ particles to two ϕ particles: $\phi\phi \rightarrow \phi\phi$. There are an infinite number of Feynman-like diagrams for this transition. Some of the simpler Feynman diagrams are:

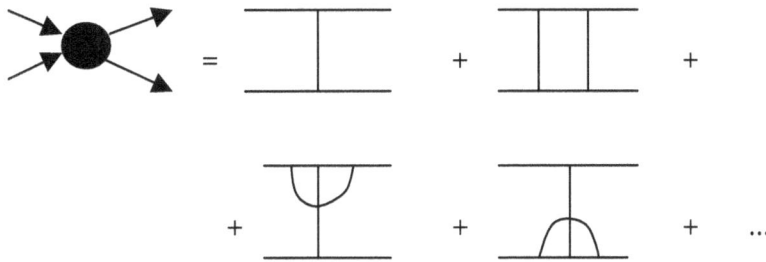

Figure 4.2.5   Diagrams for an elastic collision with two incoming particles and two outgoing particles.

Each of these Feynman diagrams corresponds to several time-ordered Feynman-like diagrams.

In evaluating these diagrams to calculate probabilities we must remember that we are ignoring space-time aspects such as particle propagators and momenta. So the calculation of the probability amplitude for this process becomes a counting problem of the number of diagrams that exist for each power of g^2. The probability amplitude for each diagram is a power of g^2.

Counting diagrams is a combinatorial mathematics problem that we will not explore in detail because it is peripheral to our interests. Consequently we will simply express the probability amplitude as:

$$A_2(g) = \sum_{n=1}^{\infty} a_n g^{2n}$$

where the mathematical expression on the right represents a sum from n = 1 to infinity and where the numbers a_n are integer numbers equal to the number of different diagrams having a power of g^{2n} as its probability amplitude. Each intersection of lines (called a vertex) contributes a factor of g to the amplitude for that diagram. The powers of g for the simpler diagrams that appear on the previous page are:

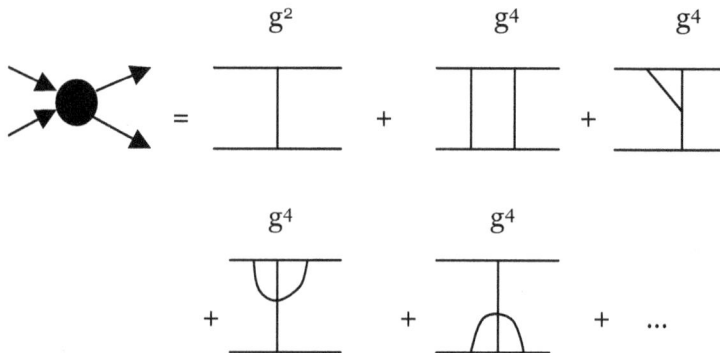

Figure 4.2.6 The power of g for some simple diagrams.

The value of the first constant a_1 is 1 since there is only one Feynman diagram with amplitude value g^2 for this process – the first diagram to the right of the = sign in Fig. 4.2.6. We will treat all time-ordered variations of a Feynman diagram as contributing one to the value of a_n. The value of a_n grows rapidly as n increases. For large n the value of a_n is of the order of $(n!)^2$. Consequently the sum is an asymptotic power series. Here again the details are not important for us. We are not looking for a numerical result.

The (unnormalized) relative probability for the transition $\phi\phi \rightarrow \phi\phi$ is:

$$P_2 = |A_2(g)|^2$$

where |...| represents the absolute value of the complex probability amplitude. In quantum theories a probability is always the square of the absolute value of the probability amplitude. The probability P_2 is a relative probability that must be normalized – multiplied by a factor that makes the sum of the probabilities of all possible outcome states equal to one. To calculate this probability we must calculate the sum of all the relative probabilities P_n to produce any number of ϕ particles from a two ϕ input state.

$$\phi\phi \rightarrow \phi \dots$$

The total of the relative probabilities is:

$$P = \sum_{n=1}^{\infty} P_n$$

where P_n is the relative probability to produce an output state with n ϕ particles.

The calculation of the relative probabilities P_n for n ϕ particles output states is similar to the calculation P_2. For example, for three particles

$$A_3(g) = \sum_{n=1}^{\infty} b_n g^{2n+1}$$

where the numbers b_n count the number of distinct diagrams with the power g^{2n+1} and

$$P_3 = |A_3(g)|^2$$

The absolute (normalized) probability to produce an n ϕ particle output state is

$$Q_n = P_n/P$$

The sum of all possible output state probabilities equals one:

$$1 = \sum_{n=1}^{\infty} Q_n$$

The modified ϕ^3 Quantum Field Theory provides a simple example of a Quantum Probabilistic Grammar™. We will now turn to the Standard Model and examine its Quantum Probabilistic Grammar™. Because it encompasses a much larger number of different particles (letters) and interactions (grammar rules) it will be significantly more complex.

4.3 Probability Amplitudes of Quantum Grammars

Quantum Grammar rules associate a probability amplitude with each production rule. To find the probability for a transition from an initial state to a final state we must calculate the probability amplitude for the transition through the repeated application of the production rules for each possible path from the initial state to the final state. We assume each initial state of the Quantum Turing machine begins with probability amplitude one. (This is a normalization condition for the initial state in reality.)

When a production rule is applied to a state to produce a transition to a new state the current probability amplitude is multiplied by the probability amplitude of the production rule. Thus the total relative probability for the passage from an initial state i to a specific final state f is

$$P_{fi} = \left| \sum_{paths} a_1\, a_2\, a_3\, \cdots\, a_{n(path)} \right|^2$$

where the sum is over all finite paths that lead from the initial state to the final state through the application of all relevant production rules, and where $a_1 a_2\, a_3\, \cdots\, a_{n(path)}$ is the product of the probability amplitudes of the production rules for each individual path. P_{fi} is similar in form to a Feynman path integral expression (Feynman(1965).

The value of n is path dependent and thus denoted n(path). The relative probability is the absolute value squared of the sum of products of the amplitudes. The relative probability must be normalized to produce an absolute probability such that the sum of the absolute probabilities of all possible final states is one:

$$P_{absolute\text{-}f\,i} = NP_{f\,i}$$

$$N = \sum_{f} P_{f\,i}$$

$$1 = \sum_{f} P_{absolute\text{-}f\,i}$$

where the sums are over all possible final states f.

Thus we have a well-defined method for calculating the probability of a transition from an initial state to a final state that is illustrated by the preceding examples. The fact that it bears some resembance to the path integral methods for quantum mechanics pioneered by Feynman (See Feynman (1965).) suggests that Quantum Grammars are of interest to physics. This author was first struck by the similarity of string transitions via production rules to path integrals in 1981. After all, a jagged (discrete) Feynman path is really a string of coordinates marking the end points of each line segment of which the path is composed. Thus each path in a Feynman sum over paths can be represented by a string. The evolution of a path from line segment to line segment can be viewed as the repeated application of a probabilistic production rule. The path sum equivalent of the probability amplitude of a production rule is an exponentiated Hamiltonian factor that is a function of the change in string coordinates "due to the production rule."

5. Standard Model Quantum Grammar

5.1 Grammar Production Rules of Quantum Electrodynamics

We will start by considering the grammar production rules of the Quantum Electrodynamics (electromagnetism) sector of the Standard Model. The production rules corresponding to the electromagnetic interaction term for electrons and positrons in the Standard Model

$$\bar{e}Ae$$

are:

<u>Electron-Positron QED Production Rules</u>

$$e \rightarrow eA$$
$$e \rightarrow Ae$$
$$eA \rightarrow e$$
$$Ae \rightarrow e$$
$$p \rightarrow pA$$
$$p \rightarrow Ap$$
$$Ap \rightarrow p$$
$$pA \rightarrow p$$
$$ep \rightarrow A$$
$$pe \rightarrow A$$
$$A \rightarrow ep$$
$$A \rightarrow pe$$

where e represents an electron, p represents a positron, and A represents a photon. The production rules describe the emission and absorption of photons by electrons and positrons as well as the annihilation of an electron and positron to produce a photon, and the decay of a photon into an electron-positron pair.

An example of an interaction between two electrons in the linguistic approach is:

$$\overset{1}{ee} \rightarrow \overset{2}{eAe} \rightarrow \overset{3}{ee}$$

where the electrons interact by exchanging one photon. One Feynman-like diagram for these transitions is:

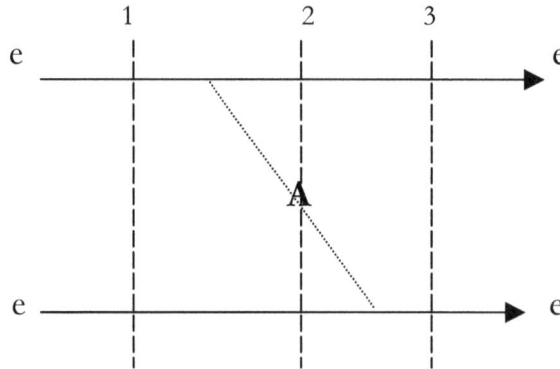

Figure 5.1.1 A diagram showing how two electrons interact by exchanging a photon. As time increases the electrons move from left to right. The upper electron emits the photon. This corresponds to the left e transitioning to eA using the grammar rule e → eA. (There is a similar diagram Fig. 5.1.2 in which the lower electron emits the photon.)

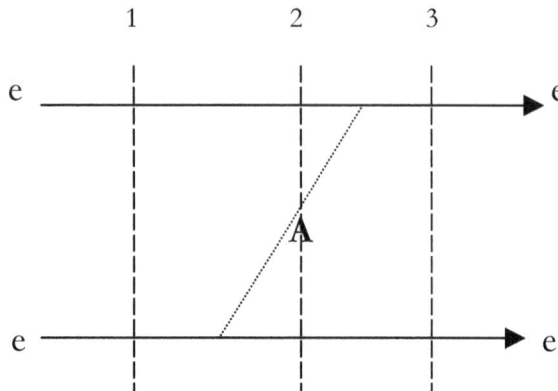

Figure 5.1.2 Another diagram with two electrons interacting by exchanging a photon. In this case the lower electron emits the photon. This corresponds to the right e in the initial ee string transitioning to Ae using the grammar rule e → Ae.

The vertical slices in the Feynman diagram which are numbered 1, 2 and 3 correspond to the three numbered strings in the transitions generated from the production rules. Each string has an ordering that corresponds to the order of particles as you descend a slice. For example slice 2 in Fig. 5.1.1 has an electron, photon, and another electron in that order as you descend corresponding to string 2 above.

Another Feynman-like diagram (fig. 5.1.2) that contributes to this process has the lower electron emitting a photon that is then absorbed by the upper electron. The Feynman diagram for this process represents the sum of both of the previous diagrams:

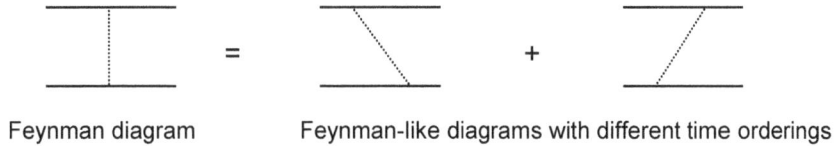

Feynman diagram   Feynman-like diagrams with different time orderings

Figure 5.1.3 A Feynman diagram represents several of our Feynman-like diagrams with different time orderings of particle emission and absorption.

A more complex example of a Feynman-like diagram appears in Fig. 5.1.4.

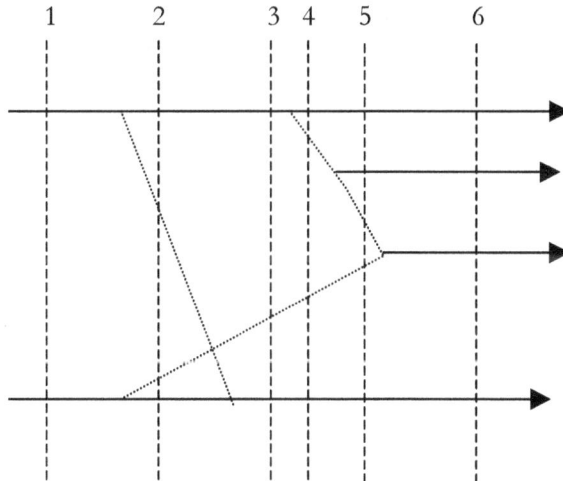

Figure 5.1.4 A diagram for the collision of two electrons that produce a new electron-positron pair: ee → eepe.

Six slices appear corresponding to the various intermediate states in this complex electron-electron interaction. The production rules can be used to generate a sequence of strings that correspond to the slices:

$$1 \qquad 2 \qquad 3 \qquad 4 \qquad 5 \qquad 6$$
$$ee \;\rightarrow\; e\Lambda\Lambda e \rightarrow e\Lambda e \rightarrow\; e\Lambda\Lambda e \;\rightarrow\; eep\Lambda e \rightarrow\; eepe$$

29

As you descend each slice the particles are ordered in the same way as the corresponding string. For example, as you descend slice 5 the order of the particles is electron, electron, positron, photon and electron, and the string is eepAe.

The transitions between character strings have an ambiguity. For example the above transition

$$\overset{2}{\text{eAAe}} \rightarrow \overset{3}{\text{eAe}}$$

could have taken place through (eA)Ae → (e)Ae with the left (upper) e absorbing an A or through eA(Ae) → eA(e) with the right (lower) e absorbing an A. (Parentheses are used for grouping to show which electron absorbed the photon.) This ambiguity reflects the fact that there are several possible time orderings. The preceding diagram Fig. 5.1.4 actually corresponds to eA(Ae) → eA(e). The right (lower) electron absorbs the photon.

To calculate the probability of an actual physical transition occurring such as ee → eepe we must take account of all diagrams with all possible time orderings for the specified input and output states. This is a monumental chore since an infinite number of diagrams are involved. Normally only the simplest diagrams are evaluated since they dominate the electromagnetic interactions of electrons and positrons.

The above examples show how the electromagnetic interaction part of the Standard Model Lagrangian can be viewed as defining a grammar. The grammar has corresponding Feynman-like diagrams. As we pointed out earlier, these types of diagrams have time orderings that are similar to the time orderings that appeared in perturbation theory calculations before 1950.

The Weak Interaction and Strong Interaction parts of the Standard Models also define grammars. Consequently we can view the complete Standard Model Lagrangian as defining a grammar where the "letters" (alphabet or vocabulary) are the elementary particles of the model and the Feynman-like diagrams corresponding to the Standard Model can be viewed as a sequence of strings generated by applying the production rules specified by the Standard Model Lagrangian.

5.2 Production Rules for the Weak and Strong Interactions

The Weak and Strong interaction terms in the Standard Model Lagrangian are also easily translated into grammar production rules (although the process is laborious since there are so many of them). We will illustrate these cases using the Weak interaction terms:

$$\bar{\nu}_e W^- e$$

and

$$\overline{\nu}_\mu W^- \mu$$

where ν_e represents an electron neutrino, ν_μ represents a muon neutrino, μ represents a muon, W^- is a gauge field of the Weak interaction and e is an electron; and the Strong interaction term

$$\overline{u}Gu$$

where u is a u quark and G represents gauge fields of the Strong interaction.

Notice that there are several types of neutrinos: electron neutrinos, muon neutrinos and tau neutrinos. The three kinds of neutrinos have different internal quantum numbers that distinguish them. Neutrinos do not have electromagnetic charge. They are neutral as their name suggests. Each kind of neutrino has a corresponding charged partner. We are familiar with the electron. The other charged partners are the muon and tau particle. These charged particles are like heavy electrons for the most part. The three charged leptons also have distinguishing internal quantum numbers.

The preceding interaction terms imply production rules such as:

$$e \rightarrow W^- \nu_e$$

$$e \rightarrow \nu_e \, W^-$$

$$W^- \rightarrow e \, \nu_e$$

$$W^- \rightarrow \nu_e \, e$$

$$\mu \rightarrow \nu_\mu \, W^-$$

$$\mu \rightarrow W^- \nu_\mu$$

$$u \rightarrow Gu'$$

$$u \rightarrow u'G$$

and so on where e is an electron, p is a positron, W^- is a negative W gauge boson, ν is a neutrino, G is a Strong interaction gauge boson and u and u' are u quarks which may have different color quantum numbers.

These production rules generate string equivalents of Weak interaction transitions such as muon decay:

$$1 \qquad 2 \qquad\quad 3$$
$$\mu \; \rightarrow \; W^- v_\mu \; \rightarrow \; e v_e v_\mu$$

which has the corresponding Feynman-like diagram:

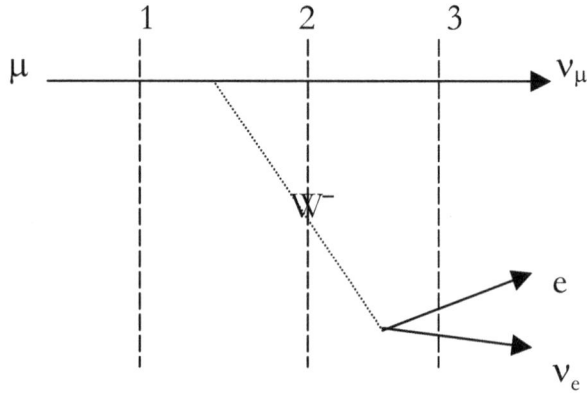

They also generate string equivalents of Strong interaction transitions between quarks and color gauge fields such as:

$$1 \qquad 2 \qquad 3$$
$$uu \rightarrow uGu \rightarrow u'u'$$

with the corresponding Feynman-like diagram:

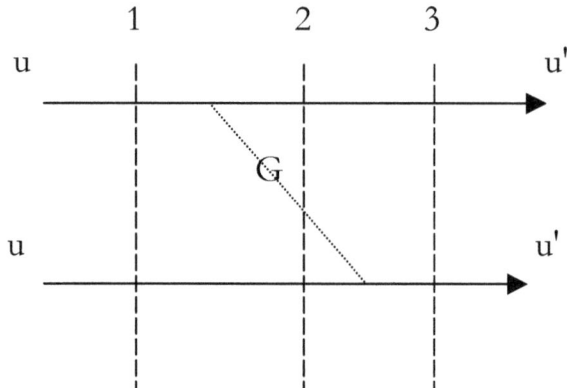

The preceding examples are among the simplest cases of the infinite variety of Feynman-like diagrams that can be generated from the Standard Model production rules.

5.3 The Standard Model Quantum Grammar™

At this point we have created a view of the Standard Model of elementary particles in which the particles are an alphabet of at least thirty-six letters (eighteen quarks, six leptons, twelve gauge fields, plus possibly other particles such as Higgs bosons).

The particle alphabet can be combined into strings that represent input or output states of scattering particles as well as bound states. Transitions between strings take place through quantum grammar production rules and correspond to time-ordered Feynman-like diagrams.

So we now have a map between the Standard Model and a language, with letters, words and a grammar;, and an interpretation of the language in terms of rules of calculation and experimental setups.

The linguistic representation of the Standard Model that we have developed omits many important calculational details as well as important properties of particles such as spin and particle momenta. These features could be added in a direct way. The focus of our investigation is on the essentials of the acts of creation and annihilation of particles in particle interactions.

The character string transition approach based on production rules is equivalent to the Feynman diagram approach. It does however provide a different, and simpler, view. Interestingly the Feynman diagram approach to calculations is very difficult and tedious – but it often leads to a simple result due to the massive cancellation of many complex terms with each other. (An attempt to simplify perturbation theory diagram calculations was made by Cheng and Wu in the early 1970's and by this author privately.) The simple linguistic approach might be a hint of a more efficient way of calculating in quantum field theories like the Standard Model where complications are absent from the very beginning.

Whether or not the linguistic approach leads to a less complicated theory or method of calculation remains to be seen. However we have now obtained a rather amazing result. After 2500 years of speculation on the nature of matter we have developed a surprisingly simple theory (for everything except gravitation) called the Standard Model that can be viewed, in part, as a quantum type 0 computer language. It has an alphabet (vocabulary) of roughly 36+ particles, and a set of production rules specified by the interaction part of the Standard Model Lagrangian.

This situation represents something of a miracle. There is no reason that Nature should have so few particles that interact with each other through a simple set of rules. A computer language theorist would call the language of the Standard Model a language with a finite representation. Simply put, this means the words of the language can be generated from a finite vocabulary (alphabet or set of particles) and a finite set of production rules.

5.4 The Standard Model Language is Surprisingly Simple

Many physicists feel that there are too many elementary particles in the Standard Model. From the point of view of a language theorist *the finite language representation of the Standard Model is a very special situation.* As Hopcroft and Ullman[3] point out, "there are many more languages than finite representations." Languages can have infinite alphabets or infinite sets of production rules or other complications.

The physical equivalent of an infinite alphabet would be a universe with an infinite number of different types of matter. Every particle of matter could have a different mass and differ in other properties. From this point of view the Standard Model is truly a marvel of simplicity.

The Standard Model, in fact, is a very compact, finite description of most of the known features of our universe. The linguistic view of the Standard Model suggests we should view elementary particles as symbols or clumps of data – a vocabulary. The interactions of the elementary particles involve the creation or annihilation of particles – creation in the deepest form seen by man.

Remarkably, the data flowing through a computer can be viewed as being transformed from one form to another and being output to different destinations. Data flowing through a computer can be divided into streams that can be sent to different output channels such as a printer or the computer screen. For example, the character, 'a', can be a data item in a stream of data that is sent to both a printer and the screen:

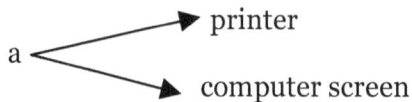

The streaming of data characters to different output channels is quite analogous to the output of particles from particle scattering as we have seen.

The simple production rules that describe particle scattering in the Standard Model suggest a fundamental simplicity at the core of Reality that goes far beyond the speculations of philosophers and scientists of earlier ages.

---

[3] J. E. Hopcroft and J. D. Ullman, *Formal Languages and Their Relation to Automata*, (Addison-Wesley, Reading, MA, 1975) page 2.

6. Quantum Turing Machines

6.1 What are Turing Machines?

The linguistic map of the Standard Model leads to a number of questions. One important question is the nature of the Turing machine that accepts this language. A Turing machine is a generalized theoretical computer that is often used to analyze computational questions in computer science.

A personal computer can be viewed as a special purpose Turing machine. A personal computer has memory in the form of RAM and hard disks. A personal computer has built-in programs that tell it what to do when data is input into the computer. Similarly a Turing machine also has memory and instructions within it telling it how to handle input and how to produce output from a given input.

When we type input on a computer keyboard or have input come from another source such as the Internet or a data file, the input has to be in a form that the computer can handle. Similarly, the input for a Turing machine must have a specific form for the Turing machine to accept it, process it and then produce output. In the case of a Turing machine we say the input must be presented in a language that the Turing machine "accepts". In this context the word "accepts" means a format that the Turing machine can recognize and analyze so that it can process the data to produce output.

The language of the Standard Model is a quantum type 0 computer language. A Chomsky type 0 language requires a Turing machine to handle its productions. Because particle transitions are quantum and because the left side of a production rule can have several possible right sides (For example, a photon can transition to an electron-positron pair, a quark-antiquark pair and so on.) the Turing machine for the Standard Model language must be a non-deterministic Quantum Turing Machine. It must accept a non-deterministic language.

6.2 Features of Normal Turing Machines

Before examining a Quantum Turing Machine for the Standard Model we will look at the features of "normal" Turing machines. A normal Turing machine consists of a finitely describable "black box." Its features are describable in a finite number of statements. But it has an infinite tape. The tape plays the role of computer memory. The tape is divided into squares. Each square contains a symbol or character. The character can be the "blank" character or a symbol. A tape contains blank characters followed by a finite string of input symbols followed by blank characters.

The black box consists of a control part and a tape head. The control part has a finite set of rules built into it (the "program") and a finite memory that it uses as a scratch pad normally. The tape head can read symbols from the tape one at a time and can move the tape to the left, right, or not move it, based on instructions from the control part.

The tape head tells the control the symbols that it is scanning from the tape and the control decides what action to have the tape perform based on the scanned symbols and information (the program and data) stored in the control's memory.

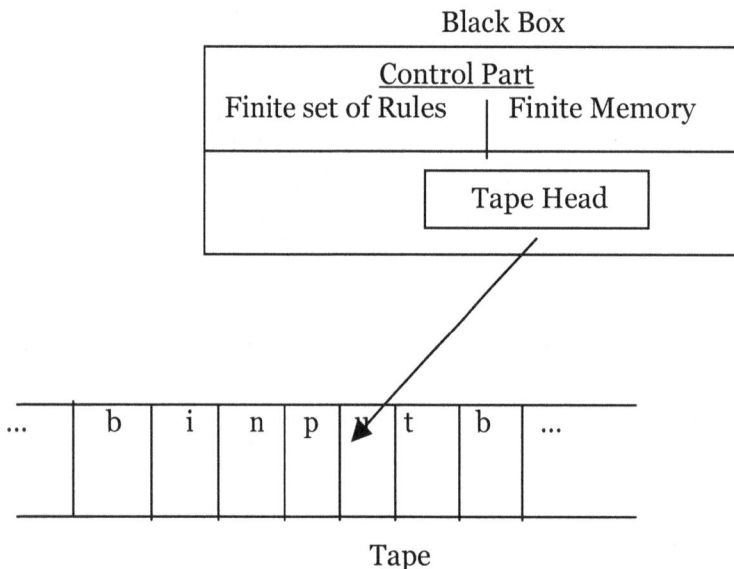

Figure 6.2.1 A schematic diagram of a Turing machine.

A set of input symbols is placed on the tape and the rules (program) in the control part are applied to produce an output set of symbols.

The Turing Machine process is analogous to elementary particle processes: an input set of particles interacts through various forces of nature and produces an output set of particles. The difference is that elementary particle processes are quantum probabilistic in nature. The laws of Physics (which appear to be finitely describable since they can be specified by the Standard Model Lagrangian plus gravity) play the role of the finite set of rules.

6.3 Quantum Probabilistic Grammars™

The major difference between Turing machine outputs, and particle physics outputs are that output states in particle physics are quantum probabilistic. A given set of input symbols (particles) can produce a variety of output states with different probabilities calculable in the Standard Model. We need a Quantum Turing Machine to handle this more complex situation.

Quantum (and non-quantum) Turing Machines can be pictured in a convenient way by viewing the control part as containing a tape on which the rules are inscribed, the current state of the Turing machine is specified, and the current symbol being scanned by the tape head is stored.

The grammar rules of a Quantum Turing Machine are quantum probabilistic. In the simplest case each grammar rule has an associated number that we will call its relative probability amplitude. We will call this type of Quantum Probabilistic Grammar™ a *factorable Quantum Probabilistic Grammar™*.

6.4 Factorable Quantum Turing Machines

A calculation of the probability for a transition in a factorable quantum Turing machine from a specified input string to a specified output string is based on the following rules:

1. The relative probability amplitude $A_{fi}(p)$ for a sequence of grammar rule transitions on a path p from the initial state i to the final state f is the product of the relative probability amplitudes a_k for each transition:

$$A_{fi}(p) = a_1\, a_2\, a_3\, \ldots\, a_{n(p)}$$

where $n(p)$ is the number of transitions for the path p.

2. The relative probability amplitude for an input string i to be transformed to a specified output string f, $A_{tot\, fi}$, is the sum of relative probability amplitudes for each possible sequence of transitions, paths, that leads from the input string to the output string.

$$A_{tot\, fi} = \sum_{paths} a_1\, a_2\, a_3\, \ldots\, a_{n(path)}$$

3. The relative probability P_{fi} for an input string i to be transformed to an output string f is the absolute value squared of the relative probability amplitude for the input string to be transformed to the output string.

Stephen Blaha

$$P_{fi} = |A_{tot\,fi}|^2 = |\sum_{paths} a_1\, a_2\, a_3\, \cdots\, a_{n(path)}|^2$$

4. The absolute probability for an input string i to be transformed to a specified output string f is the relative probability P_{fi} divided by the sum N of the relative probabilities for the input string to be transformed to all possible output strings using the quantum grammar rules. This rule guarantees that the sum of the probabilities is one.

$$P_{absolute\text{-}fi} = NP_{fi}$$

$$N = \sum_f P_{fi}$$

with the result

$$1 = \sum_f P_{absolute\text{-}fi}$$

The ϕ^3 theory example considered previously is an example of a factorable Quantum Probabilistic Grammar™ that could define the program of a factorable Quantum Turing Machine. The grammar rules for the ϕ^3 theory example were:

Transition	Relative Probability Amplitude
$\phi \rightarrow \phi\phi$	g
$\phi\phi \rightarrow \phi$	g

The relative probability amplitude for

input state: $\phi\phi$ $\rightarrow$ output state: $\phi\phi$

was shown to be

$$A_2(g) = \sum_{n=1}^{\infty} a_n\, g^{2n}$$

in the preceding chapter. The relative probability for this process is:

$$P_2 = |A_2(g)|^2$$

and the absolute probability for this process was shown in chapter 5 to be

$$Q_2 = P_2/P$$

where

$$P = \sum_{n=1}^{\infty} P_n$$

with P_n being the relative probability for a two ϕ particle input state to become an n ϕ particle output state.

6.5 Entangled Quantum Turing Machines

A general Quantum Probabilistic Grammar™ is an *entangled Quantum Probabilistic Grammar™* in which we cannot assign a probability amplitude to each individual grammar rule transition. The relative probability amplitude for a sequence of grammar rule transitions is not the product of the relative probability amplitudes of each transition.

Instead, in an entangled Quantum Probabilistic Grammar™ the relative probability amplitude for a sequence of grammar rule transitions is a function of the combined set of grammar rule transitions. Thus we call it *entangled*. (This use of the word entangled is not the same as the use of entangled in the context of entangled quantum states although its quantum nature does establish a relationship.)

The Standard Model grammar is in fact an entangled Quantum Probabilistic Grammar™ if we take account of the momenta and spins of the input and output particles. An illustration of the entanglement in the Standard Model is electron scattering through the exchange of two photons. The Feynman diagram for this process with particle momenta displayed is:

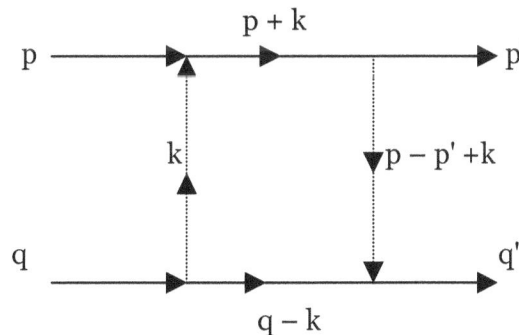

An electron with momentum p collides with an electron with momentum q and exchanges two photons (dotted lines). The momentum of each electron and photon is displayed. The outgoing electrons have the momenta p' and q'. The probability amplitude for this diagram has a factor of e^4 (a power of e – the QED coupling constant – for each of the four transitions or vertices that generate or absorb a photon). This factor is similar to the probability associated with a transition in factorable Quantum

Grammars™. *In addition* there is a factor associated with the overall pattern of transitions. The resulting probability amplitude is proportional to

$$\frac{e^4 \int d^4k \, N(k)}{k^2[(p+k)^2 - m^2] \, [(q-k)^2 - m^2] \, [(p-p'+k)^2 - m^2]}$$

where N(k) is a numerator that depends on the electron momenta, spins, and the integration "loop" momentum k. The integral depends on the details of the arrangement of the transitions and thus is an example of the entanglement of the quantum grammar. The calculation of probability amplitudes in Quantum Field theories is described in many books on the subject.

The Quantum Probabilistic Grammar of the Standard Model is an entangled Quantum Probabilistic Grammar™.

7. The Standard Model Quantum Computer

7.1 The Standard Model Quantum Turing Machine

The Quantum Turing Machine that corresponds to the Standard Model has a number of exciting features that distinguish it from conventional Turing Machines.

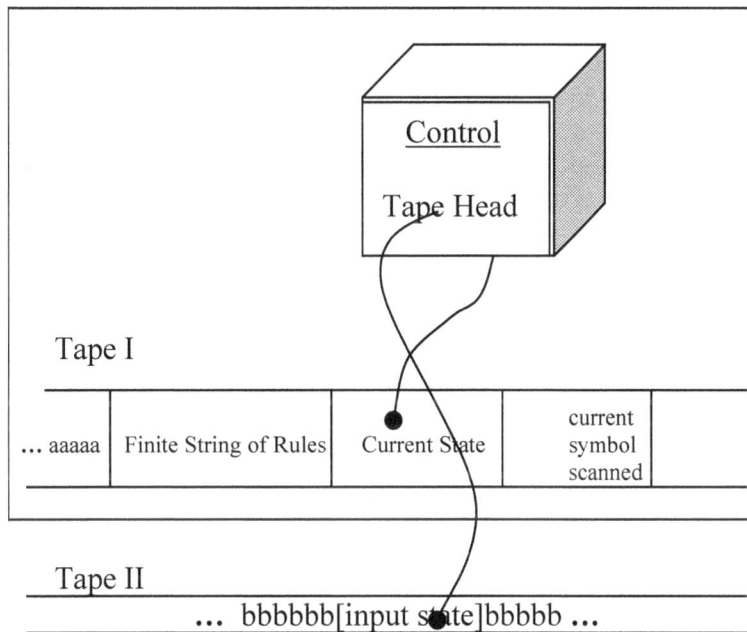

Figure 7.1.1 A Quantum Turing Machine. Tape I plays the role of the computer memory. Tape II is the memory for input and output.

First it accepts a language that has a finitely describable entangled Quantum Grammar™. Although the Standard Model has an entangled Quantum Grammar™ the grammar rules are *finitely describable*. Finitely describable means that the rules can be specified by a finite set of symbols. The rules generated from the interactions of the Standard Model are finite in number and each rule consists of a finite number of symbols. Thus the grammar rules generated from the Standard Model are finitely describable.

A Quantum Turing Machine can be visualized as consisting of a control element and two tapes that play the role of computer memory. The control has a tape head that reads and writes symbols to the two tapes.

Tape I contains the specification of the grammar rules expressed as a finite string of symbols, the current state of the Quantum Turing Machine, and other data. Tape II contains the input string. After applying the grammar rules in a quantum probabilistic way an output state is generated. The output state is placed on Tape II in the simplest Quantum Turing Machine implementation.

The behavior of a Quantum Turing Machine can be viewed as:

1. The Turing machine begins in the input state specified on tape II. An input string is placed on tape II. The other memory locations on tape II are filled with blank characters. In our case this string is a list of symbols for an input set of elementary particles that are about to interact. The Turing machine we are considering accepts any state consisting of a finite number of elementary particles. The connection between Turing machines and computer languages is brought out at this stage. A machine "accepts" a language if it can take any sentence (set of particles in our case) of the language, and perform a computation producing output (a set of output particles in our case). A Turing machine that accepts a language is an embodiment of the grammar of the language.

2. The Quantum Turing Machine applies the grammar rules to the input set of states in all possible ways to produce an output state that is a quantum superposition of states. Each possible output state has a certain probability of being produced.

3. The probability for producing a specified output state from a specified input state can be calculated as we illustrated in the simple example discussed earlier using the relative probabilities associated with the Quantum Grammar™ rules of the modified ϕ^3 theory.

4. The set of possible states of a Quantum Turing Machine[4] is infinite unlike non-Quantum Turing Machines that only have a finite number of possible states.

The Standard Model Quantum Turing Machine™ has some distinctive features:

---

[4] D. Deutsch, Proceedings of the Royal Society of London, A **400** 97 (1985) describes (universal) Quantum Computers and points out they can simulate continuous physical systems because they have a continuum (infinite number) of possible states. As page 107 points out "a quantum computer has an infinite-dimensional state space". Quantum Computers are equivalent to Quantum Turing Machines as we will see.

1. Since the order of the particles in the input state string is not physically important we will consider the input string to be actually all permutations of the order of the particles in the input.

2. Since the Turing machine is quantum, the rules are probabilistic in nature: a given set of input particles will in general produce many possible output particle states. Each output state will have a certain probability of being produced that can be calculated using the Standard Model.

3. The Quantum Grammar™ rules of the Standard Model Quantum Turing Machine™ embody internal symmetries that impact on the form of the input and output states.

4. Since the momenta and spins of the input and output particles are physically important the Standard Model Quantum Turing Machine™ must take account of these properties in the input and output states as well as internally when calculating transition probabilities.

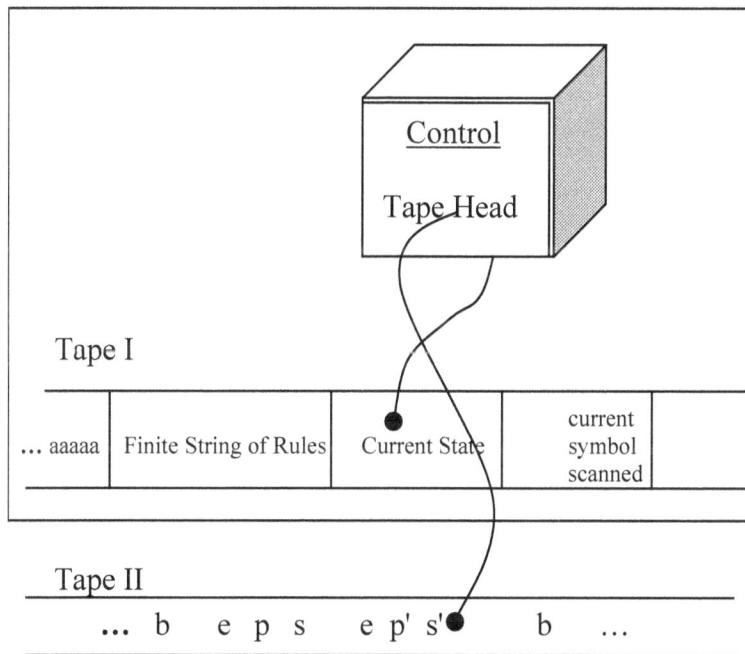

Figure 7.1.2 A Quantum Turing Machine. There is an initial state of two electrons with momenta p and p', and spins s and s' respectively. The momenta and spins are part of the specification of the input particle state.

So we must picture an input particle state on tape II as containing not only particle symbols but also momenta and spin data.

To get an idea of how a Quantum Turing Machine takes an input set of particles and produces a set of output particles we will consider the case of two electrons colliding with such energy that an electron-positron pair is created:

$$ee \rightarrow eepe$$

where e represents an electron and p represents a positron (the electron's antiparticle). One of the corresponding Feynman-like diagrams for this process is:

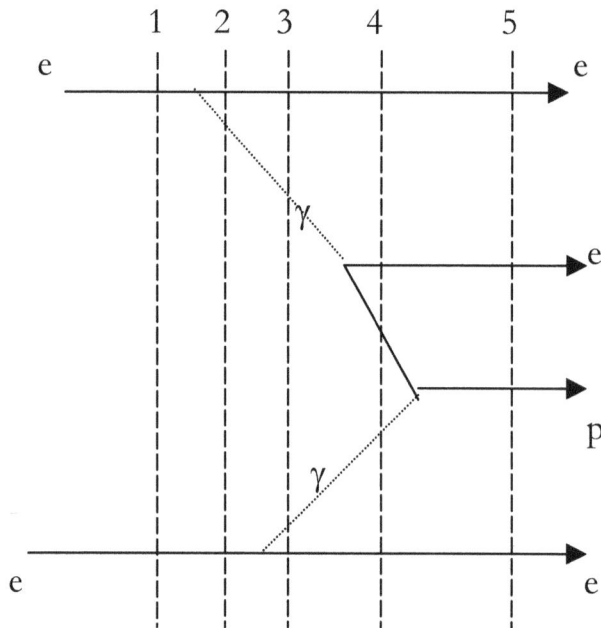

where γ represents a photon. The input string can change according to the grammar rules in the following way (There are also other ways of changing.):

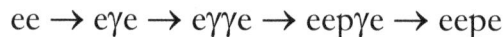

$$ee \rightarrow e\gamma e \rightarrow e\gamma\gamma e \rightarrow eep\gamma e \rightarrow eepe$$

In addition, since the Quantum Turing Machine is probabilistic there are many – in fact, an infinite number – of ways in which the transition

$$ee \rightarrow eepe$$

can take place – each with its own probability of happening. The sequence of transitions shown above is only one of these possible ways. The total probability of this transition is the square of the sum of the probability amplitudes for all possible ways according to quantum mechanics. Nature requires us to take account of all possible ways of transitioning from the input state of particles to the output state of particles.

In addition, electron-electron scattering can produce many other output states depending on the initial energy of the electrons. Each output state has its own probability of occurring. Some examples are:

$$ee \rightarrow eq\underline{q}e$$

$$ee \rightarrow eepepe$$

$$ee \rightarrow e\mu\underline{\mu}e$$

$$ee \rightarrow e\mu\underline{\mu}epe$$

where q represents a quark, μ represents the muon antiparticle and $\underline{q}$ represents an antiquark.

The Quantum Turing Machine representation does raise several interesting prospects for the theory of elementary particles embodied in the Standard Model. First, the Quantum Turing Machine representation raises the possibility that some of the powerful techniques and general results of the theory of computation can be brought over to physics and perhaps provide guidance on the next stage after the Standard Model.

Secondly, and perhaps more importantly, the separation of the input and output states (they are on tape II) from the intermediate calculational states of the Turing machine (that are on tape I) is suggestive of a somewhat different approach to the fundamentals of particle interactions. The space-time of the incoming and outgoing particles may be different from the "space-time" describing the interactions and internal structure of the interacting particles. This view is based on treating tapes I and II as representing separate space-times.

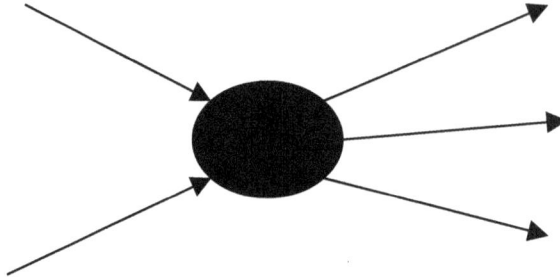

Figure 7.1.3 Two particles collide and generate a three particle outgoing state in a Quantum Field Theory.

A precursor of this point of view appears in Quantum Field Theory. In Quantum Field Theory the interaction of particles is viewed as consisting of three phases: an initial state where the particles are widely separated and distinct, an interaction region where the particles "collide" and interact perhaps creating new particles, and a final state where the outgoing particles are widely separated.

In conventional Quantum Field Theory the space-time in the interacting region is conventionally assumed to be the same as the space-time of the incoming and outgoing states. Nevertheless Quantum Field Theory distinguishes the interaction region from the region of the incoming and outgoing particles.

The SuperString approach to the theory of elementary particles introduces a separate space-time to describe the elementary particles. Elementary particles are viewed as strings vibrating in this space-time. Can one view the SuperString space-time as tape I and the external behavior of the elementary particles taking place in normal space-time as tape II? Perhaps the Quantum Turing Machine representation of the Standard Model is the key to the next level of our understanding of elementary particles and Nature.

8. Quantum Computers and Fock Space

Up to this point we have been looking at Quantum Turing Machines. Recently much excitement has been generated by Quantum Computers. A Quantum Computer is an alternate formulation of the Quantum Turing Machine concept.

In 1982 Richard Feynman[5] discussed a new theoretical concept called a Quantum Computer. Quantum Computer concepts had been developed by Benioff[16], Deutsch[16], and others in preceding years. Feynman provoked considerable interest in his view of the possible role of Quantum Computers.

A Quantum Computer can be viewed as similar to a "normal" computer in many ways. However it is not deterministic. It transitions from one state to another in a quantum probabilistic way.

Feynman's Quantum Computer was particularly adapted to simulating physical quantum systems. He gave a concrete picture of it as a space-time lattice that has two possible states at each point on the space-time lattice. See Fig. 8.1.

Feynman then hypothesized that every finite quantum mechanical system could be exactly described (exactly simulated) with such a system if the system were treated as a finite lattice of interacting spins and a suitable interaction was "chosen" between the spins at the adjacent lattice points. (We however will be interested in infinite lattices.)

A Quantum Computer starts with the spins on the lattice points that correspond to an initial quantum state of the quantum system being simulated. Then the interaction between the spins causes transitions in the spins mirroring the evolution of the quantum system being modeled. The output state of the Quantum Computer would correspond to the output state of the quantum system being simulated. In particular the probability of a transition to a specified output state of the Quantum Computer would be identical to the probability of a transition of the quantum system being modeled to its corresponding output state.

---

[5] R. P. Feynman, International Journal of Theoretical Physics, **21**, 467 (1982). Paul Benioff, Jour. Stat. Phys. **22**, 563 (1980); Jour Stat. Phys. **29**, 515 (1982); Phys. Rev. Letters **48**, 1581 (1982); Int. Jour Theoret. Phys. **21**, 177 (1982). D. Deutsch, Proceedings of the Royal Society of London, **A 400** 97 (1985) and **A 425** 73 (1989).

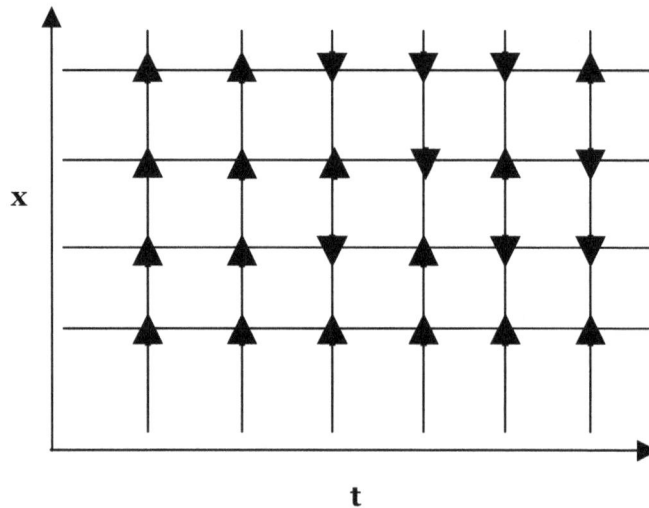

Figure 8.1 A Quantum Computer viewed as a space-time lattice of points. At each intersection point there are two possible states: occupied or unoccupied. Alternately we can use a spin ½ description and say that we have spin "up" or spin "down" at each space-time point (indicated with up arrows or down arrows). Or we can say that a computer bit exists at each space-time point that can have the value 1 or 0.

Feynman's concept of a Quantum Computer was concrete and physical – you could view it as a lattice of spins – an idea that physicists were comfortable with. Since then, numerous individuals have explored theoretical aspects of quantum computers including D. Deutsch.[6] Deutsch provided a more general formulation of Quantum Computers based on quantum bits that he called a universal quantum computer. His universal quantum computer is capable of "perfectly simulating every finite, realizable physical system." Quantum computers have been shown to offer significant advantages over normal computers in some types of calculations.[7]

The concept of the Quantum Computer as developed by Benioff, Feynman, Deutsch and others is an alternate representation of the Quantum Turing Machine concept. The Quantum Turing Machine concept is based on a linguistic approach with a

---

[6] D. Deutsch, Proceedings of the Royal Society of London, A **400** 97 (1985) and A **425** 73 (1989).
[7] P. W. Shor, Algorithms for Quantum Computation: Discrete log and Factoring, Bell Laboratories paper, 1998. Many other excellent papers on the advantages of Quantum Computers have appeared since Shor's seminal paper.

Quantum Grammar™, and a vocabulary of symbols. We choose to develop this representation of Quantum Computers because it was the natural approach for the Standard Model. The Standard Model interaction Lagrangian specified the Quantum Grammar™ with a vocabulary consisting of the fundamental particles of the Standard Model.

Can we map our Quantum Turing Machines to Quantum Computers and show they are equivalent concepts? Actually this mapping is quite easy to describe.

The map or correspondence between the Quantum Turing Machine and a Quantum Computer of the type of Feynman and Deutsch is based on the following chain of mappings:

$$\text{Symbol} \rightarrow \text{number} \rightarrow \text{bit pattern} \rightarrow \text{lattice spins}$$

Each symbol in the vocabulary of a Quantum Turing Machine can be given a numerical value just as alphabetic characters in a real computer are given numerical values internally. The numeric value of each symbol can be written, base 2, as a binary string of zeros and ones just as modern computers do. The string of zeros and ones specifies bits in a Quantum Computer. The initial quantum state of a Quantum Computer is the combined set of bits corresponding to the symbols in the input string of symbols of the corresponding Quantum Turing Machine.

The Quantum Grammar™ of the Quantum Turing Machine maps to an interaction between the spins on the lattice of the Quantum Computer. The interaction between the spins must correspond to the Quantum Grammar™ in the sense that, for any given input state, the output state of the Quantum Computer must correspond to the equivalent output state of the Quantum Turing Machine.

Another way of stating this correspondence between the Quantum Grammar™ and the lattice spin interaction is that the probability of producing a specified output state from a specified input state must be the same for a Quantum Turing Machine and its equivalent Quantum Computer for all input and output states. The Quantum Grammar™ must correspond to an equivalent interaction between the spins on the Quantum Computer lattice.

$$\text{Quantum Grammar} \leftrightarrow \text{Interaction between spins of lattice}$$

Thus we have established a mapping between Quantum Turing Machines and Quantum Computers.

8.1 The Continuum Limit of a Quantum Computer – A Superspace

A Quantum Computer has a lattice of points. At each point there is a bit that can be in an "up" or a "down" state. Let us imagine "shrinking" the lattice so the space between lattice points becomes smaller and smaller. Eventually the lattice approaches what could be called the continuum limit where the points form a continuous space. At each point of this space there is a bit that can be either "up" or "down" and, as a result, we can view this space as a *superspace* similar to that encountered in Supersymmetry theories. We will call this limiting case of the Quantum Computer a *Continuum Quantum Computer™*.

At each lattice point the spin is either up or down. As the lattice spacing shrinks we encounter discontinuities in the spin. The discontinuities form a countable set of measure zero, i.e. they can be counted using the set of integers. We smooth over these discontinuities to have smoothly varying spins in the continuum limit of the lattice. The information contained in the smoothly varying spins is the evolving data in the Quantum Computer.

In Supersymmetry theories[8] there is a parameter space consisting of space-time coordinates (parameters) which we will denote as x^μ, and Grassmann variable parameters denoted θ and $\bar{\theta}$. Grassmann parameters are not normal variables because they satisfy anti-commutation rules.

The Continuum Quantum Computer™ also has an extended space-time. It has a space-time parameter x^μ together with a spin at every space-time point. The index μ is a number specifying a coordinate. For ordinary space-time μ ranges from 0 to 3. The coordinate x^0 is the time coordinate. The coordinates x^1, x^2 and x^3 are the space coordinates.

The spin "dimension" parameter of the Continuum Quantum Computer™ can be viewed as a Grassmann variable parameter. To illustrate this correspondence let us imagine that we construct a quantum wave function $\psi(x)$ that specifies the probability amplitude to find "spin up" vs. "spin down" at each space-time point x. Let

$$\psi(x) = \begin{bmatrix} \psi_1(x) \\ \psi_2(x) \end{bmatrix}$$

be the (Weyl) spinor wave function for the quantum computer in the continuum limit. The component $\psi_1(x)$ is the probability amplitude that the spin is "up" at the point x, and the component $\psi_2(x)$ is the probability amplitude that the spin is "down" at the point x. We assume $\psi(x)$ is continuous except possibly for a set of measure 0.

$\psi(x)$ can be rewritten in terms of spinors as:

$$\psi(x) = \psi_1(x)s_1 + \psi_2(x)s_2$$

where

[8] We will follow the approach of D. Bailin and A. Love, *Supersymmetric Gauge Field Theory and String Theory* (Institute of Physics Publishing, Philadelphia, PA, 1994) pages 13, 23, and 34.

$$s_1 = \begin{bmatrix} 1 \\ 0 \end{bmatrix}$$

and

$$s_2 = \begin{bmatrix} 0 \\ 1 \end{bmatrix}$$

are spinors. Now the Grassmann variable θ has a Weyl spinor index θ_a as well. So we can represent our wave function as

$$\psi(x) = \psi_1(x)\theta_1 + \psi_2(x)\theta_2$$

Thus our Continuum Quantum Computer™ maps directly to a space of the type required for Supersymmetry.

Now we take the Supersymmetric transformations and reinterpret them in terms of the physics of the Continuum Quantum Computer™. In a Supersymmetric space the form of a Supersymmetric "rotation" is:

$$x^\mu \quad \longrightarrow \quad x^\mu + a^\mu - i\xi\sigma^\mu\overline{\theta} + i\theta\sigma^\mu\overline{\xi}$$

$$\theta \quad \longrightarrow \quad \theta + \zeta$$

$$\overline{\theta} \quad \longrightarrow \quad \overline{\theta} + \overline{\zeta}$$

where a^μ specifies a shift (translation) in space-time coordinates, and ζ and $\overline{\zeta}$ are constant anti-commuting Grassmann parameters. A Supersymmetric "rotation" can shift the space-time coordinates and inter-mix them with the Grassmann parameters. The $-i\xi\sigma^\mu\overline{\theta} + i\theta\sigma^\mu\overline{\xi}$ expression is an ordinary numeric expression – not an anti-commuting Grassmann value. The Supersymmetric "rotation" is reminiscent of the rotation between space and time that we see in the Theory of Special Relativity. Only here, in Supersymmetric space, it rotates between ordinary space-time coordinates and Grassmann coordinates.

The interpretation of the Supersymmetric "rotation" from the viewpoint of the Continuum Quantum Computer™ is quite interesting. To understand it we must first realize that the information in the memory of a Quantum Computer is not only the bits in the memory lattice. The distribution of the bits – how they change as one goes through memory is also part of the information contained in the Quantum Computer's memory. With that in mind we can view the Supersymmetric transformation that intermixes space-time and the Grassmann spinor parameters as a rotation between the data bits in the Continuum Quantum Computer™ and the space-time of the Continuum Quantum Computer™.

If the spin interactions of the Continuum Quantum Computer™ are invariant (unchanged) under a Supersymmetric transformation then the transformed Continuum Quantum Computer™ is equivalent to the original Continuum Quantum Computer™. (We assume that edge effects – effects of the edge of the Continuum Quantum Computer's space are negligible.) Invariance can be used as a form of data validation since deviations from symmetry would indicate "bad bits" of data.

8.2 SuperSymmetric Continuum Quantum Computer™

If we now imagine creating a Continuum Quantum Computer™ that is equivalent to a Quantum Turing Machine™ for an enhanced theory of elementary particles that includes SuperSymmetry, then its space will be of infinite extent. We will call this computer the SuperSymmetric Continuum Quantum Computer™. (We can also create a Continuum Quantum Computer™ for SuperString theory as we show in a later chapter.)

We can picture a SuperSymmetric Continuum Quantum Computer™ as in Fig. 8.2.1 with edges shown for the sake of illustration.

The entire universe is contained in the SuperSymmetric Continuum Quantum Computer™. The Lagrangian for that theory can be added to the "edge" of the universe as the rules (or program) part of the quantum computer forming an augmented universe.

The Lagrangian for the universe that we add to the "real" universe can be thought of as occupying a universe of its own. Spaces, or universes of Lagrangians, have been discussed by Kenneth Wilson and others. (We will discuss sets of Lagrangians in conjunction with Gödel numbers in a subsequent chapter.) An interesting question raised by the existence of a space of Lagrangians is whether a principle exists in the space of Lagrangians that can be used to deduce the Lagrangian that Nature implements in a way similar to the manner that we deduce the laws of Physics from the Standard Model Lagrangian. One possibility, that is suggested by the computer framework of this discussion, is a Lagrangian that is generated through a self-organizing process of the sort envisioned in Artificial Intelligence studies. This possibility remains to be explored.

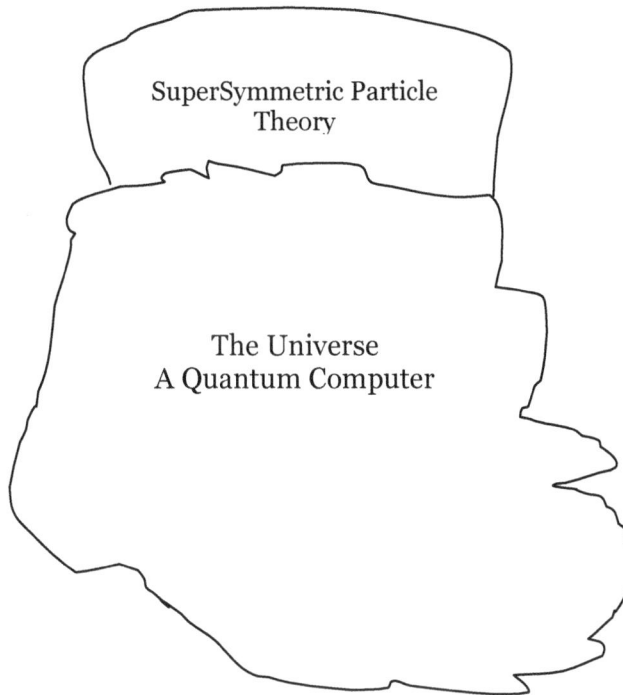

Figure 8.2.1 Visualization of a SuperSymmetric Continuum Quantum Computer™. The top part is a separate "universe" containing the Lagrangian of the SuperSymmetric Particle Theory.

Our representation of the enhanced SuperSymmetric Continuum Quantum Computer™ effectively reduces Nature to Language. Everything in the universe including the laws of Physics becomes part of a quantum computer. The space of Nature's quantum computer is filled with bit patterns describing the data and "program" or rules of the computer. And the universe evolves according to the quantum grammar of Nature.

8.3 Group Symmetries of Quantum Computers

It is interesting to note that the type of symmetries of Quantum Computers that we have been discussing, and their implications for computation, has not been addressed in the literature on Quantum Computers.

Implementing computers with symmetries introduces a form of data redundancy that can be used for data validation and data error correction. Bits that are in error and thus do not conform to the symmetry can be changed to have the correct values.

8.4 Particles and Physics Laws Become Bit Patterns

The SuperSymmetric Continuum Quantum Computer™ contains a space of bits that can describe particles. It also contains bit patterns representing the interactions of particles. Both the particles in the space and the Lagrangian possess a common symmetry. The Lagrangian is effectively the program, and space is the memory of the computer. Together they form one integral whole.

The SuperSymmetric Continuum Quantum Computer™ adds bits representing the internal symmetries to each space-time point just as the Standard Model implements internal symmetries directly without "deriving" them from a more fundamental concept such as a space-time with extra dimensions.

So we can view the SuperSymmetric Continuum Quantum Computer™ as the limit of a space-time lattice with a spin at each intersection point and a dangling thread of bits:

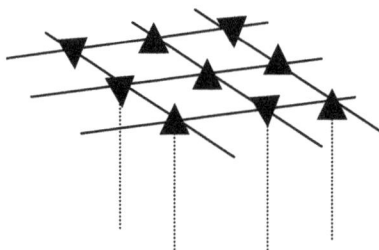

Figure 8.4.1 Lattice representation of a SuperSymmetric Continuum Quantum Computer™ with extra bits allocated for internal symmetries. The Continuum Quantum Computer™ emerges as the limit when the lattice spacing decreases to zero.

Only 6 bits are actually required for the tightest, most minimal representation of approximately 36+ particles. The actual number of bits and the role of each bit could be chosen to provide the most elegant representation of a particle's internal quantum numbers based on SuperSymmetric symmetries.

The space of the lattice has been tacitly discussed as if it were coordinate space. Actually we shall see in a few sections that the continuum limit of a Quantum Computer (where the lattice spacing shrinks to zero) can also be viewed as a second-quantized Fermi field in which the natural interpretation of the lattice space is "momentum" space. The particles' momenta and internal quantum numbers are the data resident in the Quantum Computer. The momentum distribution of each particle can be Fourier-transformed to a coordinate space representation.

8.5 Reality is Reduced to Language

Having seen particles first become symbols and then become bit patterns in a Continuum Quantum Computer™, having seen the interactions between bit patterns similarly reduced to bit patterns we now come to the conclusion that the universe and its physical laws – Reality – is in essence linguistic – Language.

8.6 The Meaning of the Standard Model Language

The purposes of human language include understanding and communication. A language also reveals something of the thinking patterns of the speakers of the language. Language not only expresses our thoughts but it also helps shape them. What can one make of the linguistic approach to the Standard Model? What does the language of the Standard Model tell us about Nature? What are its capabilities for expressing ideas? Where is the Rosetta stone that could map the language of the Standard Model to a form that we can understand in terms of its purpose and its goals? What are the reasons for its form?

The elementary particles that appear in the Standard Model, and its relatively simple form, is amazing considering the possibility of a universe with untold numbers of unrelated particles each described by a different theory. Many physicists complain the Standard Model has too many particles. But actually, as pointed out earlier, it is a small miracle that a theory containing a few particles (36+), and describable by a Lagrangian with a few parts, accounts for all known phenomena involving matter and radiation in the universe (except gravity).

This being so, one has to ask if there is a human-understandable interpretation of the theory at a level that can be explained in a few words understandable to an informed layman. (Feynman once said any good theory should have a simple idea behind it.)

John A. Wheeler developed a similar question in one of his papers[9]: "An unfamiliar computer from far away stands at the center of the exhibition hall. Some of the onlookers marvel at its unprecedented power; others gather in animated knots trying, but so far in vain, to make out its philosophy, its logic, and its architecture. *The central idea of the new device escapes them. The central idea of the universe escapes us.* [italics added]" What is the central idea behind the language of the Standard Model Quantum Computer™ and thus the material universe? Have we not found Wheeler's "unfamiliar computer from far away" to be our universe?

8.7 Deducing the Purpose of a Language from its Structure

One way to begin to develop an understanding of the central idea of the universe is to first explore the more limited possibility of *deducing the purpose of a computer language from its features*. We wish to take a language and deduce its

---

[9] J. A. Wheeler, The Computer and the Universe, International Journal of Theoretical Physics, **21**, 557 (1982).

purpose and goals from the structure and capabilities of the language. Then we might be able to apply this approach to the Standard Model computer language to develop a deeper understanding of the universe.

A reasonable starting point is to consider the set of computer languages with which we are familiar: COBOL, FORTRAN, C, C++, Java, and so on. Given any of these languages could we deduce its purpose, objectives and "philosophy"? Imagine yourself an alien and someone hands you a description of, say, the C++ language describing the features of C++ programming (written in your alien language). Could you deduce the purpose and "philosophy" of this language? An interesting question.

The C++ language is used for mathematical calculations in science, engineering and other practical applications. It is also used to manage devices such as printers, other computer devices and industrial machinery. C++ implements a "philosophy" called Object Oriented Programming. Could we have deduced this "philosophy" from the structure of the C++ language? Personally, I think not – not without serious effort – and not without an understanding of the drawbacks of the language that preceded it – the C programming language.

The C language was excellent for small scale programming projects. But because it required a programmer to track every variable and all the details of a program the C language was not appropriate for large programming projects. The C++ language was developed to handle large-scale programming. It implemented a design philosophy called Object Oriented Programming. This approach was based on grouping variables into objects. Objects were used as the central entities in programming and they were manipulated in C++ programs. By grouping many variables within an object and then manipulating the objects C++ programs are easier to create, and easier to maintain and modify.

A casual examination of C++ language features would not reveal these concepts and purposes. A detailed study of the C++ language might lead an imaginative alien to the rationale for the object-oriented features of C++.

Returning to physics, the example of the C++ language suggests that we might understand the language of the Standard Model and its "philosophy" by understanding it within the context of other languages – other physical theories, and the difficulties with other languages (theories) that the Standard Model language resolves. *This suggestion argues for a metatheory of physics theories*: a theory of the nature of physics theories and of their comparative features – a Theory of Theories.

The current situation in the development of a metatheory of physics is rudimentary. Fundamental physical theories are judged on the basis of "elegance", "beauty", "simplicity" and other ill-defined concepts. Superstring theorists have been attempting to find a theory of everything that "is the only possible, self-consistent, physically reasonable and elegant theory." They have not been successful as of yet despite twenty years of effort by a large number of extremely talented physicists. Einstein puzzled over the question whether "God had a choice in deciding the physical theory that governs the universe." The answer to this question is tied to the creation of a rigorous, well-defined metatheory of physics comparable to the metatheory of mathematics that has been developed by logicians. We will make a beginning in the metatheory of physics in chapters 13 through 16.

9. Fock Space Formalism for Continuum Quantum Computers™

A Continuum Quantum Computer™ can be viewed as defining a Fock space for two second-quantized Fermi fields, and if it is a multicephalic computer as defining a Fock space for a boson field as well. The formalism of second-quantized Fermi and boson fields can be used to describe Quantum Computers in the continuum limit. They can also approximate Quantum Computers defined by a discrete lattice.

A Quantum Computer can be described[10] as a processor consisting of a set of M 2-state (up or down) observables denoted $\{n_i\}$ where i ranges from 0 through $M - 1$, and an infinite memory consisting of a set of 2-state observables denoted $\{m_i\}$ where i ranges from $-\infty$ to ∞. The computer's memory can be viewed as either linear like a tape, or as arranged in a lattice that is numbered linearly. The tape head is assumed to be "reading" at a memory location k on the tape.

The state of a Quantum Computer can be specified with a basis of state vectors of the form:

$$| \; n_0, n_1, n_2, \ldots, n_{M-1}; \; \ldots m_{-1}, m_0, m_1, \ldots > | \; k>$$

or, in shorthand notation,

$$| \; \mathbf{n}; \mathbf{m} > | \; k>$$

The vector |k> specifies the state of the tape head. These vectors are eigenvectors of the observables and span the space of states of the Quantum Computer.

Each 2-state observable n_i or m_i can have the values 0 or 1 (bit "off" or bit "on"). Consequently, this approach is equivalent to a spin ½ picture where the values are $-\frac{1}{2}$ or $+\frac{1}{2}$.

The basis states can be easily interpreted as a set of multi-particle states in which each 2-state observable corresponds to a quantum level that can contain zero

---

[10] D. Deutsch, *Proceedings of the Royal Society of London*, A **400,** 97 (1985).

particles or one particle.[11] Thus we can view the basis of Quantum Computer states as a number representation for *fermions* (combined with the tape head state).

9.1 Processor and Memory Fermi Field

In the discussion that follows we will begin by lumping the 2-state observables of the processor together with those of the computer tape for the sake of simplicity. (We will ignore the tape head for the moment.) We will then treat these sets of observables separately.

The vacuum state for the number representation is the vector with all 2-state observable values 0:

$$\Phi_V = \mid 0, 0, 0, \ldots, 0; \ldots 0, 0, 0, \ldots >$$

or, in short,

$$\Phi_V = \mid \mathbf{0}; \mathbf{0} >$$

Creation and destruction operators can be defined that create and destroy multi-particle states (states with several one-bits). The operator $d_i^\dagger$ creates a particle in the ith quantum state or, in other words, changes the ith 2-state observable bit from zero to one.

i^{th} quantum state
(2-state observable)
↓

$$d_i^\dagger \Phi_V = \mid 0, 0, \ldots, 1, \ldots 0, 0, 0, \ldots >$$

The operator d_i "destroys" the particle in the ith quantum state or, in other words, changes the ith 2-state observable bit to zero from one. Applying a destruction operator to the vacuum state annihilates it:

$$d_i \Phi_V = 0$$

In order to guarantee the 2-state nature of the quantum levels (observables) we impose anti-commutation relations on $d_i^\dagger$ and d_i.

$$\{d_i, d_i^\dagger\} = d_i d_i^\dagger + d_i^\dagger d_i = 1$$

---

[11] See Bjorken(1965) or Weinberg(1995) for a description of fermion states or a quantum field theory texts.

$$\{ d_i, d_j \} = 0$$

$$\{ d_i^\dagger, d_i^\dagger \} = 0$$

where $\{A, B\} = AB - BA$ defines the anti-commutator of two operators.

These anticommutation relations guarantee the spectrum of each observable will be one or zero (See Bjorken(1965) or Weinberg(1995) for more details). The relation of d_i and d_j where i and j are different is not clear until we realize that a state must satisfy

$$d_i^\dagger \, d_j^\dagger \Phi_V \; = \; c d_j^\dagger \, d_i^\dagger \Phi_V$$

for $i \neq j$ where c is a phase factor ($e^{i\phi}$). The order in which bits are set cannot influence the state that is created except for an irrelevant phase factor. (It is irrelevant because probabilities are calculated from the *absolute* value squared of amplitudes causing phase factors to cancel.) The phase factor can be chosen to be −1 to obtain consistency with the anticommutation relations above. Similarly

$$d_i \, d_j \Phi \; = \; c' d_j \, d_i \Phi$$

for $i \neq j$ where c' is a phase factor ($e^{i\phi}$). The phase factor can again be chosen to be −1 to obtain consistency with the anticommutation relations above.

Combining these considerations we obtain the anticommutation relations:

$$\{ d_i, d_j^\dagger \} = \delta_{ij}$$

$$\{ d_i, d_j \} = 0$$

$$\{ d_i^\dagger, d_j^\dagger \} = 0$$

where δ_{ij} equals 1 if i = j and δ_{ij} equals zero otherwise.

The general Quantum Computer state can now be represented by a multi-particle state:

$$| \; n_0, n_1, \ldots, n_{M-1}; \; \ldots , m_{-1}, m_0, m_1, \ldots > | \; k > = N d_i^\dagger d_j^\dagger \ldots d_q^\dagger d_p^\dagger \Phi_v | k >$$

where N is a normalization constant. While it would be possible to go on and discuss normalizing states, and other issues of the number representation, we will refer the reader to Bjorken(1965) or Weinberg(1995) for further details.

The aspect that we would like to explore further is the introduction of a second-quantized Fermi field containing the creation and destruction operators that we have defined above to characterize a Quantum Computer. For this purpose we define the field

$$\psi(z, t) = \sum_i u_i(z, t) d_i$$

where the sum is over the lattice sites labeled with index number i. The Hermitean conjugate field is

$$\psi^*(z, t) = \sum_i u_i^*(z, t) d_i^{\dagger}$$

The operators ψ and ψ^* satisfy the anti-commutation relations

$$\{\psi^*(z, t), \psi^*(z', t)\} = 0$$

$$\{\psi(z, t), \psi(z', t)\} = 0$$

$$\{\psi(z, t), \psi^*(z', t)\} = \sum_i u_i(z, t) \, u_i^*(z', t) = \delta(z - z')$$

where the set of functions $u_i(z, t)$ form a complete, orthonormal set of functions. These functions should be solutions of an equation that describes the time evolution of the quantum computer (perhaps similar to the Schrödinger equation of Quantum Mechanics). They satisfy the orthonormality condition:

$$\int dz \, u_i(z, t) \, u_j^*(z, t) = \delta_{ij}$$

The parameter z can be viewed as a one-dimensional spatial parameter.

In the continuum limit the index i can be transformed into a continuous variable that can be viewed as the momentum conjugate to the z parameter.

Thus we have arrived at a second-quantized Fermi field $\psi(z, t)$ formulation that describes a Quantum Computer.

A one-particle state can be specified by:

$$\int dz\ u_i(z,\ t)\ \psi^*(z,\ t)\Phi_V = d_i^\dagger\Phi_V = \mid 0,\ 0,\ \dots,\ 0,\ \overset{i^{th}}{1},\ 0,\ \dots >$$

Multi-particle states can be constructed through a straightforward generalization.

The non-zero anti-commutation relation between ψ and ψ^* reflects a measurability issue for Quantum Computers that is familiar from the analogous phenomena in the theory of Quantum Fields. If we attempt to measure a field at two points that can be connected by a light signal then an uncertainty principle exists that prevents the exact measurement of the field at each point.[12] In the case of a Quantum Computer the measurement of a quantum bit(s) at one point in a Quantum Computer and the measurement of another bit in another part of the computer that can be connected to it by a light signal are quantum incompatible – they satisfy an uncertainty principle that prevents their simultaneous measurement.

9.2 Time Evolution of a Quantum Computer Realized with a Fermi Field

Having arrived at a second-quantized Fermi field description of a Quantum Computer it is interesting to raise the question of how it can be programmed from the perspective of this representation. Programming a Quantum Computer requires the specification of a quantum dynamics that describes the evolution of the Quantum Computer from an initial state.

There are several approaches that appear reasonable at first glance:

1. Specify a multi-particle (multi-bit) Hamiltonian using a Schrödinger-like equation that describes the time evolution of the particles (bits) of the Quantum Computer. This approach is limited by the fact that the number of particles (bits) is fixed – unlike most computations where the number of particles (on bits) varies as the computation progresses.

2. Define a second-quantized field theory for ψ in z space with interactions that specify a computation. The interactions can appear in a variety of forms. Renormalization questions (infinities that appear in computations) are not an issue since the discreteness of the Quantum computer (the lattice spacing) provides a natural cutoff eliminating problems with infinities. Interaction terms that could appear in a second-quantized field theory include:

---

[12] N. Bohr and L. Rosenfeld, Kgl. Danske Videnskab. Selskab. Mat.-Fys. Medd., **12**, 8 (1933) and Phys. Rev. **78**, 794 (1959).

$$a_1 \psi^* \psi \psi^* \psi + a_2 \psi^* \frac{\partial \psi}{\partial z} \psi^* \frac{\partial \psi}{\partial z} + \ldots$$

This approach has the difficulty that it is hard to map the interaction terms into bit operations in a direct way making it difficult to specify a program for a Quantum Computer in this manner.

3. Directly use the creation and annihilation operators to define a program for a Quantum Computer. A simple example of this approach is to consider a Quantum Computer that embodies the dynamics of an "on" bit "moving" on a tape. We assume a simple time-independent Hamiltonian implement the dynamics:

$$H = \sum_i d_{i+1}^\dagger d_i$$

Let us assume the Quantum Computer starts in a state with only the 0th bit on (or in other words a particle at the 0th lattice or tape position). Let us also define the set of one particle states:

$$\Phi_i = d_i^\dagger \Phi_V$$

Then with the time evolution operator:

$$S = e^{iHt}$$

we obtain the state of the Quantum Computer (initially in state Φ_0) at time t:

$$\Phi(t) = S\Phi_0$$

or in expanded form:

$$\Phi(t) = \sum_{n=0}^{\infty} \frac{(it)^n}{n!} \left(\sum_j d_{j+1}^\dagger d_j \right)^n \Phi_0$$

$$= \sum_{n=0}^{\infty} \frac{(it)^n}{n!} \Phi_n$$

The state of the Quantum Computer has evolved into a superposition of one-particle states residing at memory positions 0, 1, 2, ... (or alternately a superposition of Quantum Computer states with one bit "on").

The preceding example illustrates a Quantum Computer programmed to evolve dynamically according to a specified Hamiltonian. More realistic, and therefore more complex, programming can be done with this approach using creation and annihilation operators. The preceding example also illustrates the case of a nearest neighbor interaction – an interaction between adjoining bits or tape locations. The Quantum Computer state evolves into a state composed of a superposition of one "on" bit (one-particle states) states.

The preceding example can be viewed within the framework of a computer program if we assume (as usually is the case) that the execution time of the program is small $t = \delta t = n\varepsilon$ and consists of n small steps of time interval ε. Expanding the exponential to first order

$$S = e^{iHt} \simeq 1 + iHn\varepsilon$$

we see the program is

$$P = iHn\varepsilon$$

since the first term leaves the state of the computer unchanged. The change in the state Φ_0 after n program steps is

$$\Phi_c(n\varepsilon) = in\varepsilon H\Phi_0$$

Therefore it is apparent that the program consists of the same "statement" iH executed n times. A more general program can be defined by a product of "computer language" statements H_i

$$\Phi_c(n\varepsilon) = in\varepsilon H_1 H_2 H_3 \ldots H_n \Phi_0$$

which each take ε seconds to execute. Still more general quantum programs can be defined. These will be discussed in sections 12.7, 12.8 and 12.10.

9.3 Quantum Computer Memory Fermi Field

We have just seen the mapping of a Quantum Computer to a many-particle formalism, and to a second-quantized Fermi field, without differentiating between the processor and the computer memory, and neglecting the positioning of the tape head. These complications can easily be handled within the framework of the mapping.

As pointed out earlier, the general state of a Quantum Computer can be symbolized by a state of the form

$$\mid n_0, n_1, n_2, \ldots, n_{M-1}; \ldots m_{-1}, m_0, m_1, \ldots > \mid k>$$

We can define creation and annihilation operators for the memory bits m_i as before:

$$\{\, d_i, d_j^\dagger \,\} = \delta_{ij}$$

$$\{\, d_i, d_j \,\} = 0$$

$$\{\, d_i^\dagger, d_j^\dagger \,\} = 0$$

where δ_{ij} equals 1 if $i = j$ and δ_{ij} equals zero otherwise.

9.4 Quantum Computer Processor Fermi field

In addition we now define creation and annihilation operators for the processor bits which we will denote a_i having similar anticommutation relations:

$$\{\, a_i, a_j^\dagger \,\} = \delta_{ij}$$

$$\{\, a_i, a_j \,\} = 0$$

$$\{\, a_i^\dagger, a_j^\dagger \,\} = 0$$

Lastly we define a creation operator $c^\dagger$ and an annihilation operator c corresponding to the tape position observable. We assume the tape is semi-infinite – the tape locations range from 0 to $+\infty$. Therefore the spectrum of this observable – the tape read position – ranges from 0 to $+\infty$, and the creation and annihilation operators must be defined to satisfy Bose commutation relations:

$$[\, c, c^\dagger \,] = 1$$

$$[\, c, c \,] = 0$$

65

$$[c^\dagger, c^\dagger] = 0$$

With these additional operators we can now define a Quantum Computer state as:

$$\mid n_0, n_1, \ldots, n_{M-1}; \ldots m_{-1}, m_0, m_1, \ldots > \mid k> =$$

$$N \, (c^\dagger)^k a_r^\dagger \, a_s^\dagger \, \ldots \, a_t^\dagger a_u^\dagger d_i^\dagger \, d_j^\dagger \, \ldots \, d_q^\dagger \, d_p^\dagger \Phi_V$$

where Φ_V is the vacuum state of the Quantum Computer and N is a normalization constant.

Following the same strategy as earlier it is possible to define fields

$$\psi(z, t) = \sum_i u_i(z, t) \, d_i$$

and

$$\phi(z, t) = \sum_j w_j(z, t) \, a_j$$

where the sum for ψ is over the lattice sites labeled with index number i and where for ϕ is over the ϕ lattice sites labeled with index number j. The ψ and ϕ lattices are of course different. The Hermitean conjugate fields are

$$\psi^*(z, t) = \sum_i u_i^*(z, t) d_i^\dagger$$

and

$$\phi^*(z, t) = \sum_j w_j^*(z, t) a_j^\dagger$$

The operators ψ and ψ^*, and ϕ and ϕ^*, satisfy the anti-commutation relations

$$\{\psi^*(z, t), \psi^*(z', t)\} = 0$$

$$\{\psi(z, t), \psi(z', t)\} = 0$$

$$\{\psi(z, t), \psi^*(z', t)\} = \sum_i u_i(z, t) \, u_i^*(z', t) = \delta(z - z')$$

$$\{\phi^*(z, t), \phi^*(z', t)\} = 0$$

$$\{\phi\,(z, t), \phi\,(z', t)\} = 0$$

$$\{\phi\,(z, t), \phi^*(z', t)\} = \sum_j w_j z, t)\, w_j(z', t) = \delta(z - z')$$

where the set of functions $u_i(z, t)$ and $w_j(z, t)$ form complete orthonormal sets of functions. In the case of the processor bits (2-state observables or particles) the number of observables is assumed to be infinite (although the number of occupied states is finite.) These functions ψ and ϕ are solutions of equations that describe the time evolution of the quantum computer processor and tape. Normally one would expect the processor and tape states to be interrelated as they evolve with time.

The parameter z has been treated above as a one-dimensional spatial parameter. It could have been a two-dimensional, or three-dimensional, etc. parameter.

9.5 Polycephalic Quantum Computers with Tape Head Boson Field

We could have started with a multi-dimensional tape or a set of multidimensional tapes – a *Polycephalic Quantum Computer™*. The one tape head observable with creation and annihilation operators c and $c^\dagger$ cannot not be mapped to a corresponding field since there is only one observable. This situation can however be changed by introducing an array (lattice) of tape heads. Classical polycephalic Turing machines with multiple tape heads have been discussed in the literature. These Turing machines have a finite number of tape heads. We can extend the number of tape heads to be infinite (or very large) in number (perhaps form a multidimensional lattice) in the case of Quantum Computers. This extension enables us to introduce a second-quantized boson field corresponding to the lattice of tape heads:

$$\sigma(z, t) = \sum_j s_j(z, t)c_j$$

$$\sigma^*(z, t) = \sum_j s_j^*(z, t)c_j^\dagger$$

with c_j corresponding to the jth tape head. The boson field operators σ and σ^* satisfy the commutation relations

$$[\sigma^*(z, t), \sigma^*(z', t)] = 0$$

$$[\sigma\ (z, t), \sigma\ (z', t)] = 0$$

$$[\sigma\ (z, t), \sigma^*(z', t)] = \sum_{j} s_j(z, t)\ s_j^*(z', t) = \delta(z - z')$$

The appearance of both Fermi and Bose fields:

Field	Type	Role
$\psi(x, t)$	Fermion	Processor
$\phi\ (y, t)$	Fermion	Memory
$\sigma\ (z, t)$	Boson	Tapes Positions

in Quantum Computers, and the apparent similarity of the lattice space of the Quantum Computer memory to a Supersymmetric space, leads us to consider the possibility that a Quantum Computer formalism may provide a foundation for the development of SuperString implementations. This possibility will be considered later.

10. A SuperString Quantum Turing Machine

10.1 The Ideas of SuperString Theory

Among the best attempts to develop a deeper theory than the Standard Model is called SuperString Theory (although it is really an infinite set of possible theories). These theoris attempt to explain the nature of elementary particles (What is a symbol?) and how elementary particles interact with each other (What is the grammar?) based on a deep picture consisting of strings vibrating in a space with many more dimensions than the four space-time dimensions with which we are familiar.

The linguistic representation of the Standard Model described in an earlier chapter is an interesting preliminary to SuperString Theory. We will look at SuperString Theory from a Quantum Grammar™ and Quantum Turing Machine perspective in this chapter.

There is no connection between the *strings* of particles (which are actually words made of symbols) in our linguistic representation of the Standard Model, and the strings of SuperString Theories. The only thing these approaches have in common is that they are both one-dimensional constructs. A string of symbols is one-dimensional – the symbols are ordered in a row. A string in SuperString Theory is one-dimensional – it is a mathematical construct that is analogous to a string made of thread.

A string in the linguistic representation of the Standard Model is a discrete list of individual particle symbols (characters). A string in SuperString Theory is a continuous, one-dimensional space (like a thread) representing one particle. It can be visualized as a rubber band (closed string representation of a particle) or as a piece of cotton thread with two ends (open string representation of a particle).

Particles in the Standard Model are point-like. SuperString Theory treats each "fundamental" particle – quarks, electrons, and so on – as an extended string in space-time. SuperStrings are thought to be so small that their string-like nature cannot be verified experimentally with today's particle accelerators. Current experiments see particles as more or less "point-like" surrounded by "clouds of virtual particles." If experiments could probe more closely they might see particles as strings. Strings are thought to be approximately 100 billion billion times smaller than a proton.

●

Figure 10.1 Particles are point-like in the Standard Model.

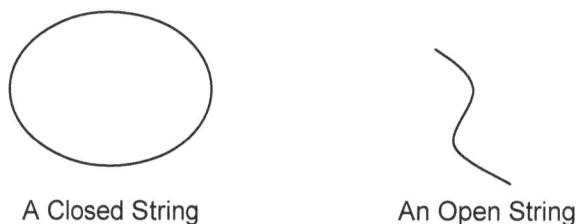

A Closed String   An Open String

Figure 10.2 Particles are strings in SuperString Theory.

In SuperString Theories elementary particles are various "vibrations" of superstrings. A superstring can vibrate in a number of different ways just like a violin string can vibrate at different frequencies. Each vibration of a superstring corresponds to a different particle. SuperString theory should have a vibration for each different type of elementary particle including gravitons – the particle that carries the force of gravity. Actually SuperString theories predict many more particles than the particles that we have actually found in nature. "Eliminating" these extra particles at current experimental energies – often by suggesting that they are too massive to be observed by current particle accelerators – is a major task of SuperString theory.

Some versions of String Theory assume a particle is a closed superstring. This type of superstring is often implemented in a form called a *heterotic* string. A heterotic string consists of two strands. In some versions of SuperString Theory one strand is a 10-dimensional string. The other strand is a 26-dimensional string. This composite string has properties that may be a closer approximation to reality than other types of strings (although there is a hope that the various types of theories are essentially manifestations of the same underlying theory).

SuperString theories have many interesting features that are described in popular[13] and technical[14] books. Perhaps the most interesting feature of SuperString theories is their high degree of symmetry. Many versions have the rotational symmetry of 26-dimension space-time. They often have conformal symmetry – a symmetry that can crudely be described as stretchability – a string can be stretched or deformed in mathematical ways. They can also have a peculiar symmetry called Supersymmetry. Supersymmetry is based on the rotation of fermions (spin ½ particles, spin 3/2 particles, and so on) and bosons (spin 0 particles, spin 1 particles, and so on) into each other.

Before Supersymmetry, fermions were fermions and had special properties that distinguished them. For example, they had Fermi-Dirac statistics based on the Pauli

---

[13] Michio Kaku, *Hyperspace*, (Bantam, Doubleday, Dell Publishing, New York, 1994) among others.
[14] Joseph Polchinski, *String Theory* (Cambridge University Press, New York, 1998). D. Bailin and A. Love, *Supersymmetric Gauge Field Theory and String Theory* (Institute of Physics Publishing, Philadelphia, PA, 1994) among others.

Exclusion Principle that stated that no two fermions could have exactly the same set of quantum numbers as we saw in the preceding chapter.

Before Supersymmetry, bosons were bosons and they had their own special features. Bosons had Bose statistics that stated that any number of bosons could have the same set of quantum numbers. These differences in statistics and other features gave bosons and fermions radically different properties.

Supersymmetry enables rotations to take place in a mathematical space of bosons and fermions. Fermions and bosons can be rotated into each other. As a result of Supersymmetry fermions and bosons are interrelated and form families. This symmetry puts important restrictions on theories that support it.

At the moment no solid experimental evidence exists that supports the idea that Supersymmetry is a symmetry of Nature. However the Supersymmetry concept is very appealing theoretically and it is an integral part of SuperString theories.

Having seen the basic ideas of SuperString theories we now try to relate them to the linguistic representation of the Standard Model. Is SuperString theory interpretable as a computer language?

10.2 A Linguistic Model of SuperString Theory?

If we take the idea of a computer language description of Nature seriously, then we have to understand how the features of SuperString theory can be viewed as composing a language.

We have seen that we can view the Standard Model as a computer language. Is SuperString theory possibly a more fundamental (lower level) language? Computer languages can have a hierarchical relation. For example, the C++ language is built on the more fundamental C language, which, in turn, is built on the even more fundamental Assembly language which, in turn, is built on machine language. Is the Supersymmetry language the machine language upon which the Standard Model language is based? The hallmark of these lower level computer languages is that they are increasingly simpler as we descend the hierarchy to the lowest level.

The lowest level computer language, machine language, has few constructs in it and directly tells the computer chips what to do. On the other hand, the interpretation of the constructs of a higher level language in terms of lower level language constructs is often complex.

10.3 A SuperString Theory Language and Turing Machine

The natural choice for a grammar for a SuperString Theory has letters (terminal symbols) that are superstrings. Each letter represents a specific string vibration. The problem with this approach is the infinite number of possible string vibrations. We can partially resolve this problem by taking all vibrations below a certain energy thus giving us a finite alphabet. Physically this is not an unreasonable approximation. However conceptually it would have been more desirable to have an alphabet consisting of all possible vibrations.

The production rules for the SuperString grammar are simple at the symbolic language level. The details of the numeric calculations behind the grammar rules are complex. They are described in books such as those listed in the References section.

A set of grammar rules for the linguistic representation of SuperString interactions is:

Rule 1. ◎ → ◎ ◎

Rule 2. ◎◎ → ◎

Rule 3. ◎◎ → ◎′ ◎′

The quote marks on the right side of rule 3 indicate that the symbols on the right side represent, in general, different particles (terminal symbols) from the particles on the left side. This notation is part of the formalism of computer language grammar rules. The grammar rules can be represented pictorially as closed string diagrams:

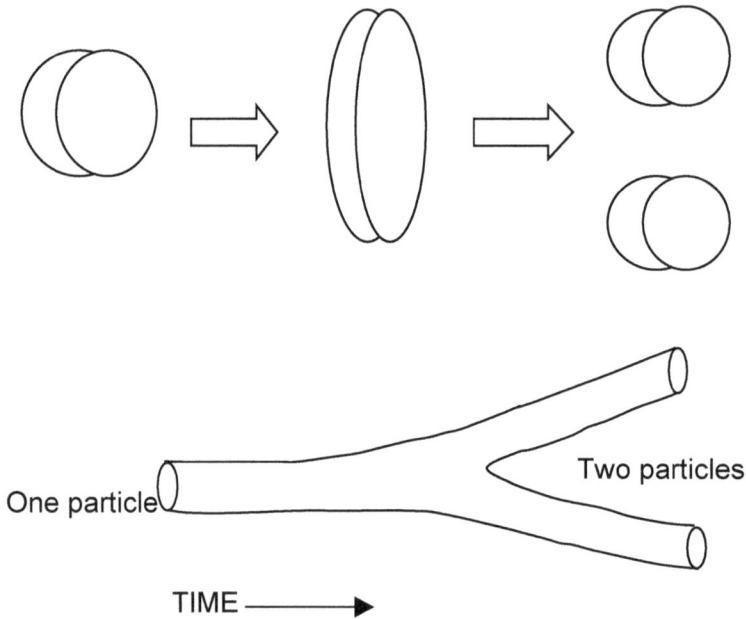

Figure 10.3.1 Two visualizations of rule 1 for the "fission" of a superstring into two superstrings. Time increases as the strings move to the right in the diagrams. The second visualization of superstrings moving in such a way as to sweep out tubes in time is the standard visualization.

The "fission" of a photon string into an electron string and a positron string is an example of this grammar rule. The strings look like tubes or pipes as they move in time: if you take many pictures of a moving string and superimpose the pictures the motion of the strings will trace out tubes "in time."

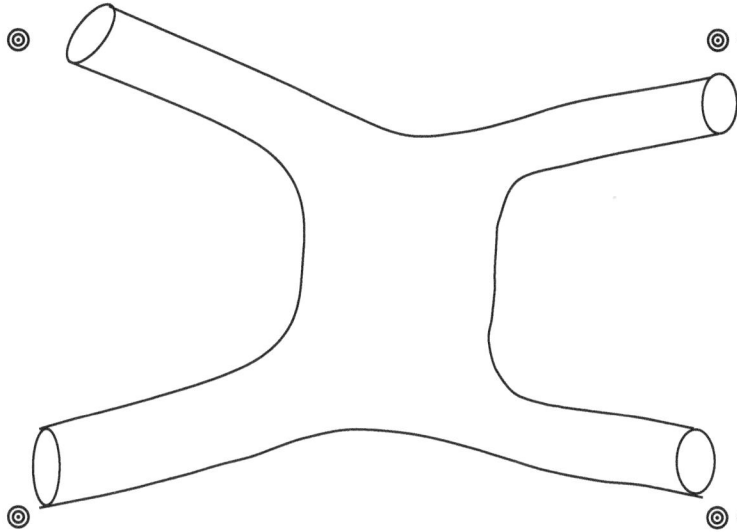

Figure 10.3.2 A diagrammatic visualization of rule 3. The superstrings look like tubes or pipes as they move in time. In this case two superstrings interact and then produce two output superstrings.

The grammar production rules for SuperStrings are much simpler than the grammar production rules for the Standard Model. This was to be expected if the SuperString Theory is in some sense a deeper theory corresponding to a lower level computer language – perhaps the equivalent of machine language for particles.

The Quantum Computer that corresponds to SuperString theory accepts input states consisting of a word containing a character string of particle superstrings. It processes the input and generates an output word consisting of a symbol string consisting of particle superstrings. Each possible output word has a corresponding probability of being produced since the Quantum Computer is probabilistic. Inside the Quantum Computer the input particles interact – producing a variety of intermediate particle states or words. Eventually the output particles are generated.

The general features of the SuperString Quantum Turing Machine™ (SQTM) are similar to the Quantum Turing Machine for the Standard Model. However the language and the theory behind the language are different. The SQTM generates intermediate superstring states based on the details of the SuperString theory that it implements. An example of an interaction between superstrings with interesting intermediate states is:

$$\circledcirc\circledcirc \rightarrow \circledcirc \, ' \, \circledcirc \, ' \, \circledcirc \rightarrow \circledcirc \, ' \, \circledcirc \, ' \, \circledcirc \, ' \, \circledcirc \, ' \rightarrow \circledcirc \, ' \, \circledcirc \, ' \, \circledcirc \, ' \rightarrow \circledcirc \, ' \quad \circledcirc \, '$$

The intermediate superstrings are generated using the three SuperString grammar rules stated earlier. This set of transitions can also be depicted with a SuperString theory diagram with time orderings:

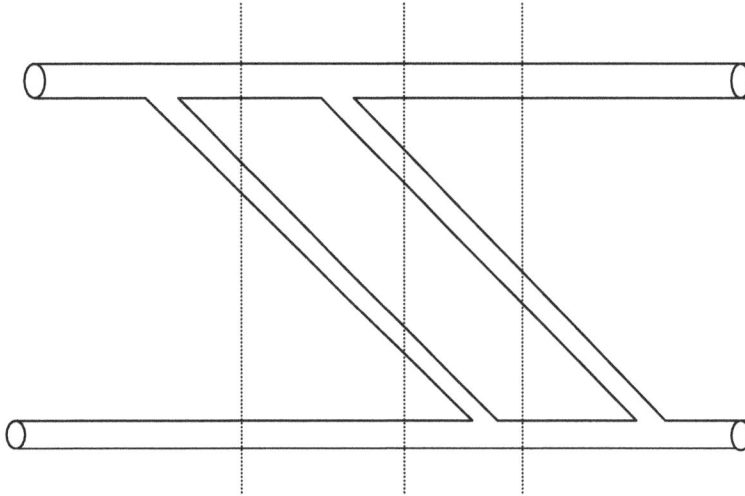

Figure 10.3.3 String diagram of two interacting superstrings that generate two superstrings in intermediate states. These generated superstrings are absorbed so that only two superstrings emerge in the output. The intermediate state "words" are indicated by vertical dashed lines. Many other diagrams contribute to this superstring-superstring interaction when realistic calculations are made.

The SQTM offers a new view of the features of SuperString Theory. This view may be helpful in deepening our understanding of the implications of SuperString Theory and its relation to the Standard Model.

10.4 Beyond SuperStrings – Hidden Dimensions as Quantum Computer Tapes

SuperString theories assume the existence of hidden dimensions that are the origin of the internal symmetries of elementary particles. These dimensions are somehow compactified (curled up). As a result we only experience the normal four dimensions of space and time.

The Quantum Computer representations of the Standard Model and Supersymmetry that we have developed suggest an alternate possibility: the internal tapes of the Quantum Computer may define a different space from the space of the input and output particles.

A Quantum Computer has several tapes. These tapes play the role of computer memory just like computer tapes. One tape – the input tape – has the input particles on it. Another tape – the output tape – has the output particles on it. (The input and output tapes could be the same tape.)

The Quantum Computer also has an internal tape that can be viewed as a set of tapes or as a multidimensional tape upon which intermediate states of the particles can be stored. We can view a Quantum Computer that processes an input state to produce an output state as:

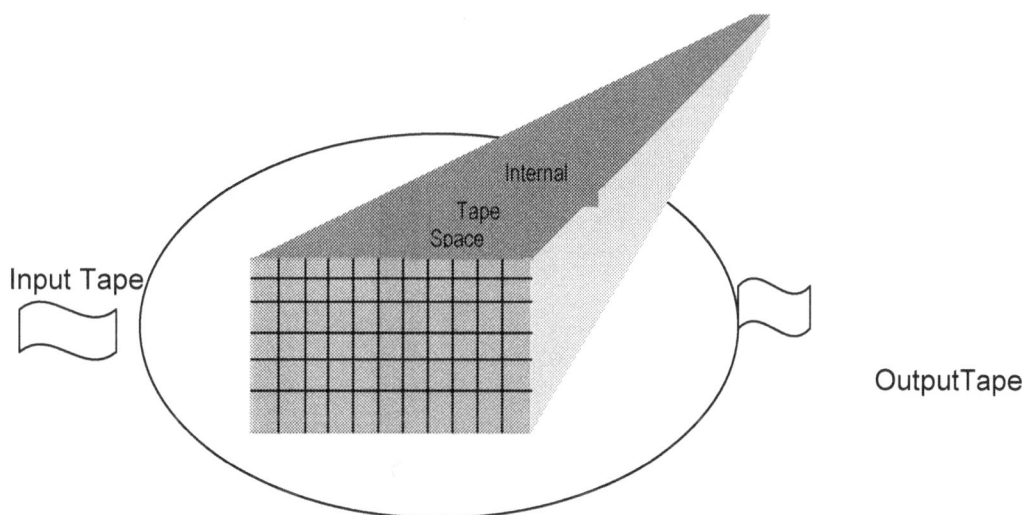

Figure 10.4.1 A Quantum Computer with an internal multidimensional space (tape).

The internal Turing machine tape defines an internal "space" to which we can assign dimensions. This space, as it is normally viewed in Turing machines, is divided into cells. However in Quantum Computers it can be made into a continuous space. Further we can make this space the space of the internal symmetries of the Standard Model or Supersymmetry Theory (or in part the space of symmetries of Supersymmetry Theory).

This concept leads us to view the space of internal symmetries as part of the Quantum Computer. This space is relevant when particles interact (collide). The internal space is not relevant and is not "visible" when particles are not interacting. This simple view agrees with our experience. The "space" of the input and output tapes interfaces with (or melds into) the internal space when particles interact. So we can expect that the internal space somehow meshes the input and output space in a smooth way.

The smoothness (continuity) of the transition between the internal space of the Quantum Computer and the external space of the input and output states could result in a unified theory that unites gravity and the internal particle symmetries. This approach is similar in flavor to the standard SuperString approach for the unification of the forces of nature.

10.4.1 Quark Confinement as Nonterminals in a Particle Theory Quantum Turing Machine

An interesting point related to the Quantum Computer representation of the Standard Model and Supersymmetry is quark confinement. Individual, isolated quarks have not been found in Nature. Quarks always appear bound together with other quarks: three quarks in a proton, a quark-antiquark pair in a pion, and so on. The inability to split a proton or any other particle containing quarks into individual quarks is an interesting phenomenon. Most theorists believe it is a feature of the Strong Interaction between quarks. Unlike other forces such as gravity the Strong Interaction becomes stronger and stronger as quarks move further apart preventing them from separating. This feature of the Strong Interaction can be implemented within the grammar of the Standard Model or Supersymmetry by making the quarks into *non-terminal symbols*.

A non-terminal symbol is a symbol that can appear within the grammar of a language but cannot be an input symbol or an output symbol. It only appears as an intermediate symbol within the grammar rules. Protons, pions and other composite particles composed of quarks can be viewed as *terminal symbols* – symbols that can appear as input symbols or output symbols. Quarks only appear inside the Quantum Computer as particles in intermediate states generated by the grammar. They are reassembled into composite particles that are output (terminal) symbols.

The Quantum Computer representations of the Standard Model and Supersymmetry provide an alternate view of elementary particles that is both simple and yet capable of embodying significant features of these theories in a straightforward way. Quark confinement is one Standard Model feature that is easily represented in the linguistic representation. The existence of hidden dimensions is also easily represented in the Quantum Computer approach as an internal multidimensional tape. The linguistic approach to elementary particle theories naturally accommodates major features of elementary particle phenomena.

11. Quantum Computers as the Foundation of SuperStrings

11.1 Introduction

Many physicists feel that SuperString Theory is the next step beyond the Standard Model. This theory is based on an interesting idea: *build matter out of the fabric of space and time*. Not the space and time with which we are familiar but rather a space and time of many more dimensions. Most of the dimensions of this higher dimensional universe "curl up" into a "tight ball" so we only see the four dimensions of space and time of everyday experience. The curled extra dimensions are thought to be the origin of the internal symmetries that particles possess in the Standard Model.

The reduction of matter to a feature of space-time is an important concept. It leads us to *an insubstantial universe consisting only of space and time that is based on quantum probabilities*. It lacks the solidity of our world of ordinary perceptions. A world governed by a language and consisting of "nothingness." Matter being mere wrinkles in space and time.

Actually space and time – since they are represented by the gravitational field which itself is a SuperString construct – also become constructs. In the end we wind up with "nothing" – a "nothing" that develops structure through the SuperString Theory. The structure "becomes real" through the existence of superstrings – particles represented by mathematical entities. "Normal" space and time now exist through the presence of gravitons – particles carrying the gravitational force – which are in reality mathematical strings. Thus Nature is reduced to mathematical form.

In the previous chapter we saw the appearance of both Fermi and Bose fields in Quantum Computers. We also saw the apparent similarity of the lattice space of the Quantum Computer memory to a space that could support Supersymmetry. These observations lead us to consider the possibility that a Quantum Computer formalism may provide a foundation for SuperString Theory. In this chapter we will consider a simple Superstring model implemented as a quantum computer.

SuperString theory today has some notable successes. However it is still a patchwork quilt of clever ideas that lead to something approaching a realistic theory of elementary particles. The situation is reminiscent of the Bohr model of the atom. It blended together disparate ideas (that could be viewed as incompatible in their origin) to produce a model of the Hydrogen atom that explained some major experimental features of Hydrogen. It was not until Quantum Mechanics developed about ten years later that a logically satisfactory theory of Hydrogen was created.

SuperString theory may be a "Bohr model" of elementary particles in wait for the right theory. The SuperString Quantum Computer™ that will be developed in this chapter offers a new representation of SuperString theory.

11.2 The Bosonic String Part of a Quantum Computer

The beginning point of our discussion is the standard view of a Quantum Computer augmented to contain a set of tapes for memory and a set of linear one-dimensional arrays of tape heads. Each array has an infinite number of tape heads and looks like a "tape" (or string) of tape heads. Therefore we will call an infinite linear array of tape heads a *string of tape heads*.

Following a similar path to the discussion in the preceding chapter we associate Fermi creation and annihilation operators with each tape location (or 2-state observable), and Bose creation and annihilation operators with each tape head. We use the phrases annihilation operator and destruction operator interchangeably.

In section 9.5 we defined an infinite set of tape heads with corresponding operators c_n with n ranging from $-\infty$ to ∞. We can redefine these operators[15] to adapt them to the development of a SuperString formalism. Let us define

$$a_n = c_n \qquad n \geq 0$$
$$a_n^\dagger = c_n^\dagger \qquad n \geq 0$$

$$\underline{a}_{-n} = c_n \qquad n < 0$$
$$\underline{a}_{-n}^\dagger = c_n^\dagger \qquad n < 0$$

The operators a_n and $\underline{a}_n$ and their hermitean conjugates can be generalized to multiple sets of tape heads (dimensions) by adding an index specifying the tape head set (dimension).

---

[15] We will use a notation consistent with standard texts on SuperStrings: D. Bailin and A. Love, *Supersymmetric Gauge Field Theory and String Theory* (Institute of Physics Publishing, Philadelphia, PA, 1994) page 157; Joseph Polchinski, *String Theory* (Cambridge University Press, New York, 1998)

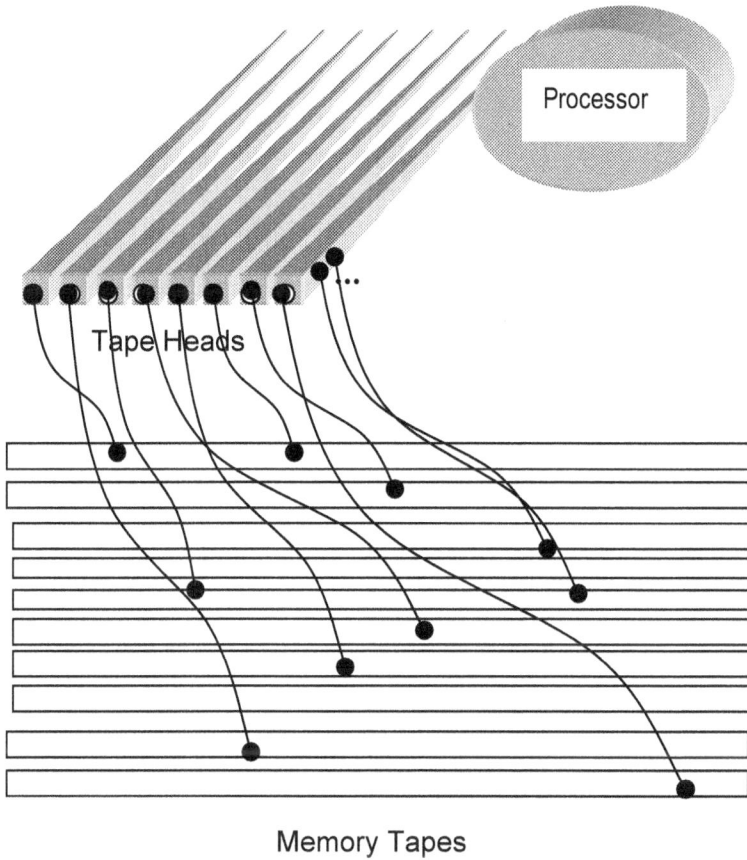

Figure 11.2.1 A Polycephalic Quantum Computer™ with infinite arrays of tape heads and many memory tapes.

The boson operators for D strings of tape heads are:

$$a_n^{\mu} \quad \text{and} \quad \underline{a}_n^{\mu}$$

with n ranging from 0 to ∞. The index μ specifies the tape head set (dimension) and ranges from 0 to D.

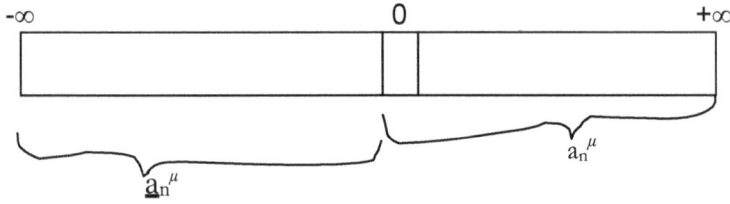

Figure 11.2.2 The μth set of tape heads. There is a tape head in the set for each value of n with operators a_n^μ for $n \geq 0$ and $\underline{a}_{-n}^\mu$ for $n < 0$.

These operators have the same tape head commutation relations seen in the previous chapter. Taking account of the added index the commutation relations become:

$$[\, a_i^\mu, \, a_j^{v\dagger} \,] = -\delta_{ij}\eta^{\mu v}$$

$$[\, a_i^\mu, \, a_j^v \,] = 0$$

$$[\, a_i^{\mu\dagger}, \, a_j^{v\dagger} \,] = 0$$

where $\eta^{\mu v}$ is zero if $\mu \neq v$ and has absolute value 1 if $\mu = v$. The tape head operators for different tape head sets (dimensions) commute with each other. The operators $\underline{a}_i^\mu$ and $\underline{a}_i^{v\dagger}$ have similar commutation relations.

To establish a connection with the SuperString literature and to make a connection with the bosonic strings of Physics we introduce the operators a_n^μ and $\underline{a}_n^\mu$ with n ranging from $-\infty$ to ∞. The index μ again specifies the tape head set (which we will call dimension hereafter). These operators are related to a_n^μ and $\underline{a}_n^\mu$ via:

$$a_n^\mu = \sqrt{n}\, a_n^\mu \qquad\qquad a_{-n}^\mu = \sqrt{n}\, a_n^{\mu\dagger} \qquad\qquad \text{for } n > 0$$

and

$$\underline{a}_n^\mu = \sqrt{n}\, \underline{a}_n^\mu \qquad\qquad \underline{a}_{-n}^\mu = \sqrt{n}\, \underline{a}_n^{\mu\dagger} \qquad\qquad \text{for } n > 0$$

These operators are fully equivalent to the oscillator coefficient operators in the mode expansion of a closed bosonic string:

$$X^\mu = x^\mu + l^2 p^\mu \tau + .5/\Sigma_{n \neq 0} (\, a_n^\mu\, e^{-2in(\tau - \sigma)} + \underline{a}_n^\mu\, e^{-2in(\tau + \sigma)} \,)/n$$

We have thus effectively created a field just as we did in the previous chapter – although with the complications of an extra index. The coordinate X^μ is a coordinate in a D dimensional space. The parameters τ and σ can be viewed as parametrizing a closed string in D dimensional space with a time-like coordinate τ and a space-like coordinate σ. For fixed τ, X^μ traces out a curve (a string) as σ varies. The terms in the summation are for a "right mover" part that depends on $\tau - \sigma$ and uses the operators a_n^μ, and for a "left mover" part that depends on $\tau + \sigma$ and uses the operators $\underline{a}_n^\mu$. The left mover modes are used to construct heterotic strings – the type of string that often appears in SuperString Theories.

The mode expansion of X^μ can be viewed as the solution of the one-dimensional wave equation:

$$\left(\frac{\partial^2}{\partial\tau^2} - \frac{\partial^2}{\partial\sigma^2} \right) X^\mu = 0$$

Introducing a metric $h_{\alpha\beta}(\tau, \sigma)$ in addition to the metric $\eta^{\mu\nu} = \mathrm{diag}(1,-1)$ that appeared earlier in the commutation relations we can develop the mode expansion for X^μ from the action for the relativistic bosonic string

$$S = \frac{-T}{2}\int d\tau \int d\sigma \, (-\det h)^{\frac{1}{2}} \, h^{\alpha\beta}\eta^{\mu\nu}\partial_\alpha X_\mu \partial_\beta X_\nu$$

by obtaining the Euler-Lagrange equations of motion, and using reparametrization invariance and conformal invariance to reduce the metric $h_{\alpha\beta}$ to

$$h_{\alpha\beta} = \eta_{\alpha\beta}$$

which then leads to the above wave equation.

The other features of bosonic strings (such as the Virasoro algebra) can be obtained by consistently applying the standard field theoretic formalism. (See Bailin(1994).)

As a result of this development we can see the tape heads of a Quantum Computer supplemented by reparametrization and conformal invariance provide a representation of the bosonic string concept. Consistency requirements select the dimension D of the bosonic string space to be 26. Thus there are 26 infinite "strings" of tape heads in the Quantum Computer for SuperString theory.

Thus we have a new concrete representation of bosonic strings. In the next section we will see the memory tapes of a Quantum Computer can be mapped to SuperStrings.

11.3 The SuperString Part of a Quantum Computer

In the preceding section we saw how arrays (strings) of tape heads could be related to bosonic strings. In this section we show the Fermi operators of memory tapes can be related to SuperStrings. In the next section we join the bosonic string discussion together with the results of this section to show how heterotic strings can naturally emerge from the parts of a Quantum Computer.

Chapter 9 showed that each tape position could be associated with creation and annihilation operators b_i and $b_j^\dagger$ satisfying the anticommutation relations:

$$\{ b_i, b_j^\dagger \} = \delta_{ij}$$

$$\{ b_i, b_j \} = 0$$

$$\{ b_i^\dagger, b_j^\dagger \} = 0$$

where δ_{ij} equals 1 if i = j and δ_{ij} equals zero otherwise. The indices i and j are integers that label tape positions and range between $-\infty$ and $+\infty$.

Let us now consider the Polycephalic Quantum Computer™ depicted at the beginning of this chapter that has a number of memory tapes. Let us suppose there are D tapes (dimensions) that are labeled with the index μ.

In order to make our discussion similar in form to the Physics literature (Bailin (1994) and Polchinski(1998) and references therein) we now define operators for Ramond boundary conditions:

$$d_n = b_n \qquad n \geq 0$$
$$d_{-n} = b_n^\dagger \qquad n \geq 0$$

$$\underline{d}_{-n} = b_{n-1} \qquad n \leq 0$$
$$\underline{d}_{-n} = b_{n-1}^\dagger \qquad n \leq 0$$

These new operators satisfy

$$\{ d_i, d_j \} = \delta_{i+j,0}$$

$$\{ \underline{d}_i, \underline{d}_j \} = \delta_{i+j,0}$$

It is easy to generalize the above anticommutation relations to:

$$\{ d_i^\mu, d_j^\nu \} = -\delta_{i+j,0}\eta^{\mu\nu}$$
$$\{ \underline{d}_i^\mu, \underline{d}_j^\nu \} = -\delta_{i+j,0}\eta^{\mu\nu}$$

where i and j range from $-\infty$ to ∞.

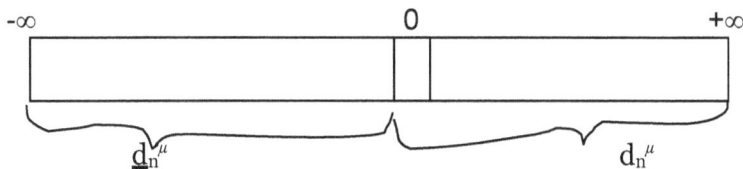

Figure 11.3.1 The μth memory tape string. There is a location on the string for each value of n with corresponding operators d_n^μ for $n \geq 0$ and $\underline{d}_{-n}^\mu$ for $n < 0$.

We can now form the mode expansions for a fermionic *right mover*, closed SuperString for periodic (Ramond) boundary conditions:

$$\Psi_R^\mu = \sum_{n \in Z} d_n^\mu \, e^{-2in(\tau - \sigma)}$$

and for anti-periodic (Neveu-Schwartz) boundary conditions:

$$\Psi_R^\mu = \sum_{n \in Z + 1\backslash 2} b_n^\mu \, e^{-2in(\tau - \sigma)}$$

where the sum extends from $-\infty$ to ∞ and where the operators b_m^μ are the corresponding Neveu-Schwartz operators with m half-integral.

The operators b_m^μ are not the operators b_n defined previously. The operators b_n^μ and $\underline{b}_n^\mu$ satisfy similar anticommutation relations to d_n^μ and $\underline{d}_n^\mu$ (see below). They are an alternate set of creation and annihilation operators for the memory tapes labeled with half integer values.

We can now form the mode expansions for a fermionic *left mover*, closed SuperString assuming periodic (Ramond) boundary conditions:

$$\Psi_L^\mu = \sum_{n \in Z} \underline{d}_n^\mu \, e^{-2in(\tau + \sigma)}$$

and anti-periodic (Neveu-Schwartz) boundary conditions:

$$\Psi_L{}^\mu = \sum_{n \in Z + 1\backslash 2} \underline{b}_n{}^\mu \, e^{-2in(\tau + \sigma)}$$

where the sum extends from $-\infty$ to ∞. These mode expansions satisfy the equations:

$$\left(\frac{\partial}{\partial \tau} + \frac{\partial}{\partial \sigma} \right) \Psi_R{}^\mu = 0$$

$$\left(\frac{\partial}{\partial \tau} - \frac{\partial}{\partial \sigma} \right) \Psi_L{}^\mu = 0$$

and the canonical anti-commutation relations:

$$\{\Psi_R{}^\mu(\tau, \sigma), \Psi_R{}^\upsilon(\tau, \sigma')\} = \{\Psi_L{}^\mu(\tau, \sigma), \Psi_L{}^\upsilon(\tau, \sigma')\}$$

$$= -2\pi\delta(\sigma - \sigma')\eta^{\mu\upsilon}$$

$$\{\Psi_R{}^\mu(\tau, \sigma), \Psi_L{}^\upsilon(\tau, \sigma')\} = 0$$

If we form the spinor

$$\Psi^\mu(\tau, \sigma) = \begin{pmatrix} \Psi_R{}^\mu(\tau, \sigma) \\ \Psi_L{}^\mu(\tau, \sigma) \end{pmatrix}$$

then we can define an action that leads to the above differential equations through a canonical approach from the Euler-Lagrange equations of motion in the covariant gauge (Bailin(1994)). The combined action for SuperStrings (SuperSymmetry requires a bosonic string term) is

$$S_0 = -\frac{1}{2\pi} \int d^2\sigma \, (-\det h)^{1/2} \, (h^{\alpha\beta} \partial_\alpha X^\mu \, \partial_\beta X_\mu + i\Psi^\mu \rho^\alpha \, \partial_\alpha \Psi_\mu)$$

where

$$\rho^0 = \begin{bmatrix} 0 & -i \\ i & 0 \end{bmatrix} \qquad \rho^1 = \begin{bmatrix} 0 & i \\ i & 0 \end{bmatrix}$$

Consequently the entire formalism of SuperStrings can be "reverse engineered" from tapes in a polycephalic Quantum Computer.

Consistency requirements select the dimension D of the SuperString space to be 10. Thus there are 10 infinite memory tapes, or memory "strings", in the Quantum Computer for SuperString theory.

The SuperString action also provides a 10-dimensional bosonic string sector. The memory of the Quantum Computer could be augmented to support an additional 10 tapes with bosonic observables (not 2-state observables) supporting a spectrum consisting of the positive integers. This addition would provide consistency with the action.

The next section joins bosonic strings and fermionic Superstrings to create a heterotic string. Heterotic string theories appear to offer among the closest connections to reality within the context of SuperString Theories as we currently know it. Other SuperString theories can also be implemented following the strategy outlined herein.

11.4 The Quantum Computer as the Foundation of Heterotic Strings

In the preceding two sections we have shown how to formulate a Quantum Computer equivalent to Superstring theory consisting of bosonic strings (from tape heads) and fermionic SuperStrings (from memory tapes). In this section we will see how to build a heterotic string within the framework of the SuperString Quantum Computer™.

The SuperString Quantum Computer™ has ten infinite memory tapes (dimensions) with fermionic operators that support a 10-dimensional SuperString space with superstrings. It also has twenty-six infinite arrays or strings of tape heads with boson operators that support a 26-dimensional space for bosonic strings.

A heterotic string is a closed string created by using right movers of a (type II) SuperString and the left movers of a bosonic string. The space of states of heterotic strings is the direct product of the spaces of the right and left movers.

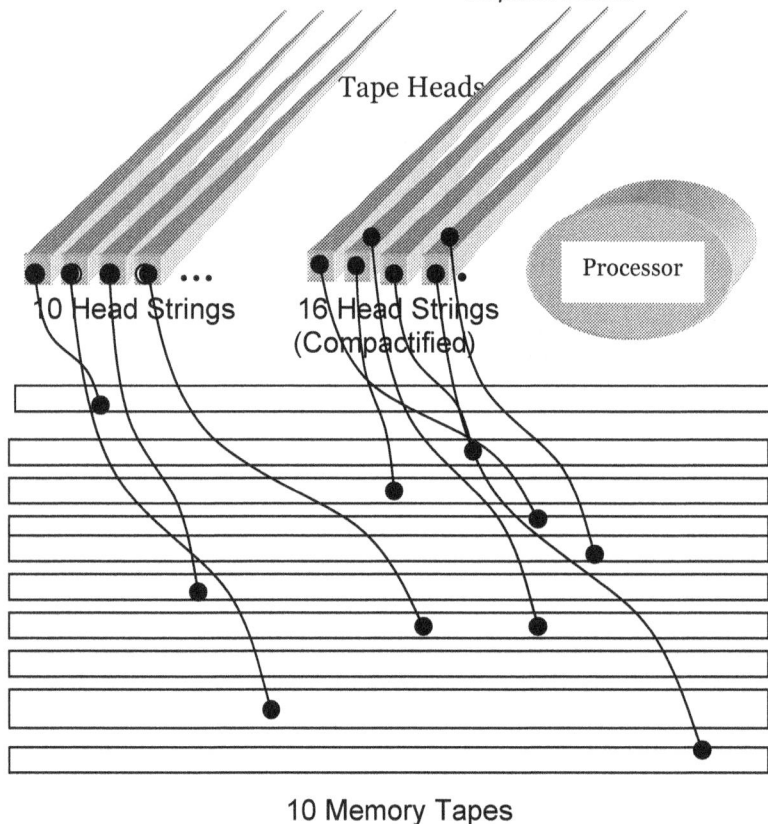

Figure 11.4.1 SuperString Quantum Computer™ with 26 infinite arrays (strings) of tape heads and 10 memory tapes.

There is an apparent problem because the space of the left movers (bosonic strings) has 26 space-time dimensions while the space of the right movers (SuperStrings) has 10 space-time dimensions. This apparent problem is normally rectified by compactifying 16 of the left mover dimensions by associating them with a 16-dimensional torus with extremely small radii of the order of the fundamental string length. So 16 dimensions shrink to "almost nothing" leaving 10 "real" dimensions for the left mover. These extra dimensions generate the internal symmetries of the elementary particles (as found in the Standard Model) through the Kaluza-Klein mechanism. In the Kaluza-Klein mechanism a higher dimensional theory containing gravity generates Yang-Mills-like gauge fields when the number of dimensions is reduced through compactification.

The creation and annihilation operators of the tape heads and memory tapes can be used to create states corresponding to elementary particles. For example, the combined tape head-memory location state

$$b^i_{-1/2}|0>_R \; \underline{a}^j_{-1}|0>_L \qquad i, j = 1, \ldots 8$$

can be decomposed into a traceless symmetric 10-dimensional graviton, an anti-symmetric tensor and a scalar dilaton. The state is defined using $|0>_R$ as the vacuum state (empty state) for right movers and $|0>_L$ as the vacuum state (empty state) for left movers. Then $b^i_{-1/2}|0>_R$ is a right mover superstring of the Neveu-Schwartz type, and $\underline{a}^j_{-1}|0>_L$ is a bosonic left mover superstring.

Thus the combination of tape heads and memory tapes supports the creation of heterotic SuperString particle states.

The next issue that the SuperString Quantum Computer™ must address is interactions between the SuperStrings. SuperString theory does not have an intrinsic well-defined way of specifying interactions between SuperStrings. So we cannot look for strong guidance from SuperString theory when we try to develop interactions within the framework of the SuperString Quantum Computer™.

SuperString Theory does have a principle that is used to specify interactions – an isomorphism between states and vertex operators. This principle is a feature of Conformal Field Theories. (SuperString theories are Conformal Field Theories.)

The vertex operators of SuperString theories are used to define interactions between SuperStrings. A vertex operator is necessary whenever SuperStrings "intersect" – when a SuperString fissions into two SuperStrings or two SuperStrings amalgamate to form one SuperString.

Since we know how to define SuperString states using the raising (creation) and lowering (annihilation) operators of the SuperString Quantum Computer™ we can use that information and the isomorphism (correspondence) to define vertex operators within the SuperString Quantum Computer™.

In view of its role, the SuperString Quantum Computer™ processor appears to be the natural part of the Quantum Computer to specify SuperString interactions. The processor is supposed to contain the program of the computer and the initial state of the computer – the input.

An alternative method to specifying interactions using the processor is to use a standard vertex operator expressed in terms of SuperString raising and lowering operators. An example is the normal ordered vertex operator:

$$:e^{ik \bullet X}:$$

where k^μ is the momentum of a SuperString particle and X^μ is the SuperString coordinate operator. The representation of the vertex operator within the quantum computer is:

$$:e^{ik\bullet X}: \rightarrow \quad \sum_{\alpha,\beta} |\alpha><\alpha|:e^{ik\bullet X}:|\beta><\beta|$$

where the sum over α and β is the sum over all possible states of the SuperString Quantum Computer. This representation of the vertex together with a perturbative expansion procedure is equivalent to a processor specification. A direct specification in terms of the processor would appear to be quite complex and physically unilluminating. It is possible that the complex representation of the vertex operator within the processor may be a hint that a different type of vertex operator is needed. However the successful use of this vertex operator in deriving the Veneziano-Suzuki formula suggests it may physically be the correct choice within the class of SuperString theories considered here.

11.5 Is the SuperString Quantum Computer™ More Fundamental than Coordinate Space SuperString Theory?

The view that we have espoused in this discussion is that SuperString Theories can be formulated in terms of Quantum Computers. The raising and lowering operators are the key features of the formalism. If one looks at the development of SuperString theories then it is very evident that the *detailed* study of many SuperString features is based on raising (creation) and lowering (annihilation) operator expressions.

In the standard development of SuperString Theory raising and lowering operators are "derived" from a space-time formalism filled with nice symmetries. In our approach the raising and lowering operators of the SuperString Quantum Computer™ are the fundamental constructs and the space-time formalism is secondary.

SuperString Quantum Computer ➡ Space-Time Theory

Establishing a Quantum Computer foundation for SuperString Theory changes the perspective from a space-time formalism to a more fundamental computational formalism. SuperStrings are freed from the trappings of space-time!

One might argue that space-time approaches to the development of physical theories have been the only successful approach in the past. But one might also argue that a space-time approach is meaningless at ultra-short distances where space and time itself are in question. The Quantum Computer formulation avoids this issue by basing SuperString theory on computation.

On the other hand, one could argue that a space-time approach should not be ruled out just because the distances are ultra-short. It works well at the larger distances with which we are familiar. Perhaps viewing short distance physics as different because we can't "see" at that level is the subtle head of anthropomorphism rearing up. Why shouldn't the space-time approach work equally as well at the short distances that we cannot directly perceive as the larger distances that we can directly perceive?

There is one outstanding example where the extrapolation from common experience to shorter distances has abjectly failed: quantum phenomena. The theories of mechanics and electromagnetism work beautifully at the level of everyday experience. We needed a major change – quantum theory - to develop a theory that worked at the sub-atomic short distance scale. Is the foundation of SuperString theory a new level with Computational Quantization?

Size Scale	Physics Theory
The Universe	SuperClassical?? OR Classical
From Galaxies to the Earth Scales	Classical
Subatomic Phenomena	Quantum Mechanics
Elementary Particles	Quantum Field Theory
Short Distance Elementary Particles	SuperStrings
Ultra-Short – Beyond Space-Time	Quantum Computation
Essence?	??

12. Quantum Computer Processor Operations and Quantum Computer Languages

12.1 Introduction

A natural question that arises when one considers Quantum Computers is the role of the Quantum Computer processor and the operations it supports. A further question of some interest is whether a "quantum computer machine language" exists and what its nature might be. Lastly the question of higher level quantum languages is also relevant. Can we develop a Quantum Assembly Language™? What is the nature of High Level Quantum Languages™? Are there, for example, equivalents to the C, C++, Java and other languages? We will address these questions in this chapter and define various levels of quantum computer languages.

12.2 Computer Machine and Assembly Languages

A traditional (non-quantum) computer can be viewed simply as a main memory, an accumulator or register (modern computers have many registers), and a central processing unit (CPU) that executes a program (instructions) step by step. It can be visualized as:

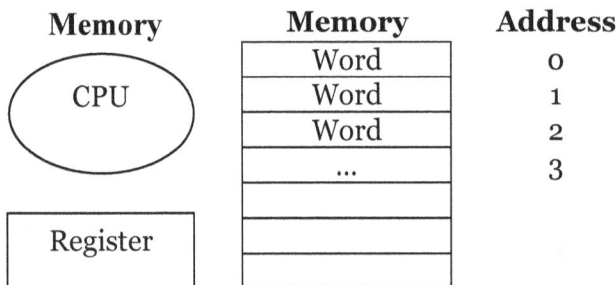

Memory	Memory	Address
	Word	0
CPU	Word	1
	Word	2
	...	3
Register		

Figure 12.2.1 A simplified model of a normal computer.

A set of data and a program (or set of instructions) is stored in memory and the CPU executes the program step by step using the data to produce an output set of data.

The basic instructions of assembly language and machine language move data values between memory and the register (or registers), manipulate the data value in the register, and implement basic arithmetic and logical operations[16]:

LOAD M – load the value at memory location M into the register

STORE M – store the value in the register at memory location M

SHIFT k – shift the value in the register by k bits (there are 8 bits in a byte. A bit can have the value 0 or 1.) This operation moves the bits in the register as seen later.)

The following arithmetic instructions modify the value in the register. The AND, OR and NOT instructions perform bit-wise and, or and not operations.

ADD M – add the value at memory location M to the value in the register

SUBTRACT M – subtract the value at memory location M from the value in the register

MULTIPLY M – multiply the value in the register by the value at memory location M

DIVIDE M – divide the value in the register by the value at memory location M

AND M – change the value in the register by "and-ing" it with the value at memory location M (described later)

OR M – change the value in the register by "or-ing" it with the value at memory location M (described later)

NOT – change the value in the register by "not-ing" it

The following instructions implement input and output of data values.

INPUT M – input a value storing it at memory location M

OUTPUT M – output the value at memory location M

---

[16] See for example Maly(1978) Chapter 8.

A computer has another register called the Program Counter. The value in the program counter is the memory location of the next instruction to execute. The following instructions support non-sequential *flow of control* in a program. A program can "leap" from one instruction in a program to another instruction many steps away (these instructions implement if-then statements in a program) and resume normal sequential execution of instructions.

TRA M   – set the value of the program counter to the value at memory location M

TZR M   – set the value of the program counter to the value at memory location M. If the value in the register is zero.

HALT   – stop execution of the program

The above set of instructions constitute an extremely simple assembly language. They also are in a one-to-one correspondence with machine instructions (machine language). Most current assembly and machine languages have a much more extensive set of instructions.

12.3 Algebraic Representation of Assembly Languages

The normal view of assembly language is that it has a word or instruction oriented format. Some assembly language programmers would even say that assembly language is somewhat English-like in part.

Computer languages in general have tended to become more English-like in recent years in an attempt to make them easier for programmers. Some view a form of highly structured English to be a goal for computer programming languages.

In this section we follow the opposite course and show that computer languages can be reduced to an algebraic representation. By algebraic we mean that the computer language can be represented with operator expressions using operators that have an algebra similar to that of the raising and lowering operators seen earlier. We will develop the algebraic representation for the case of the simple assembly language described in the previous section. There are a number of reasons why this reduction is interesting:

1. It may help to understand SuperString dynamics more deeply (later in this chapter).

2. It will deepen our understanding of computer languages.

3. It provides a basis for the better understanding of Quantum Computers and quantum computer languages.

4. It may have a role in research on one of the major questions of computer science: proving a program actually does what it is designed to do. Algebraic formalisms are generally easier to prove theorems then English-like formalisms.

5. It may be of value in the computerization of metatheories.

An algebraic representation can be defined at the level of individual bits based on anti-commuting Fermi operators. But it seems more appropriate to develop a representation for "words" consisting of some number of bits.

An algebraic representation for a word-based assembly language can be developed using commuting bosonic raising and lowering operators.

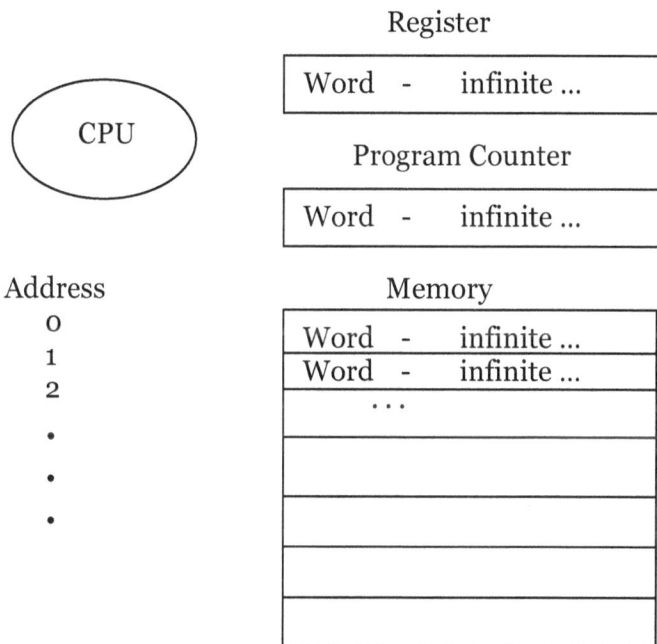

Figure 12.3.1 Visualization of a Computer with infinite words.

A word consists of a number of bits. In currently popular computers the word size is 32 bits (32-bit computer). 64-bit word computers are starting to appear. The size of the word determines the largest and smallest integer value that can be stored in the word. The largest integer that can be stored in a 32-bit word is

4,294,967,294 and the smallest integer that can be stored in a 32-bit word is 0 if we treat words as holding unsigned integers (positive integers only).

To develop a simple algebraic representation of assembly language we will assume the size of a word is so large that it can be viewed as infinite to good approximation. As a result memory locations can contain non-negative integers of arbitrarily large value. (It is also possible to develop algebraic representations for finite word sizes.) Initially we will consider deterministic computers and then proceed to quantum computers with quantum computer languages.

To establish the algebraic representation of a computer we associate a boson raising operator $a_i^\dagger$ and a lowering operator a_i with each memory location. These operators satisfy the commutation relations:

$$[\, a_i, a_j^\dagger\,] = \delta_{ij}$$
$$[\, a_i, a_j\,] = 0$$
$$[\, a_i^\dagger, a_j^\dagger\,] = 0$$

where δ_{ij} is 1 if $i = j$ and zero otherwise, and where i, j = 0, 1, 2, ... The raising and lowering operators for the register are a_0, and $a_0^\dagger$.

The ground state of the computer is the state with the values at all memory locations set to zero. It is represented by the vector

$$|\, 0, 0, 0, \ldots > \equiv |\, 0 > \equiv \Phi_V$$

A state of the computer will be represented by a vector of the form

$$|\, n, m, p, \ldots > \;=\; N\, (r^\dagger)^n (a_0^\dagger)^m\, (a_1^\dagger)^p \ldots |\, 0 >$$

where N is a normalization constant and with the first number in $|...>$ being the value in the register, the second number the value at memory location 0, the third number the value at memory location 1, and so on.

For simplicity we will defer consideration of superpositions of computer states to later in this chapter. We will first consider programs that cause a pure initial state to evolve to a pure final state. (A pure state is a state that is an eigenvector of all number operators $N_i = a_i a_i^\dagger$. for i = 0, 1, 2, ...) Within this limitation, we can set a computer state to have certain initial values in memory and then have it evolve by executing a "program" to a final computer state with a different set of computer values in memory. The "program" is a mapping of the instructions of an assembly language program to algebraic expressions in the raising and lowering operators.

12.4 Basic Operators of the Algebraic Representation

We assume that we are dealing with positive integers only in the register and memory of the computer (called *unsigned integers*). (It is possible to generalize this example to *signed integers* (positive and negative).) The key operators that are required for the algebraic representation are:

Fetch the Value at Memory Location m (using the mth Number Operator)

$$N_m = a_m{}^\dagger a_m$$

For example,

$$N_m \mid \ldots, n, \ldots > = n \mid \ldots, n, \ldots >$$

m^{th} memory location value

Set the Value at Memory Location m to Zero

$$M_m = \frac{(a_m)^{N_m}}{\sqrt{N_m!}}$$

The above expression for M_m is symbolic. The expression represents the following expression in which the operators are ordered to avoid complications (c-numbers etc.) that would appear in a different order of factors. The expression follows from expressing $(a_m)^{N_m}$ as an exponential and expanding the exponential with each term ordered as shown:

$$M_m \equiv \sum_q \frac{(\ln a_m)^q N_m{}^q}{q!} \frac{1}{\sqrt{N_m!}}$$

where the sum ranges from 0 to ∞. When M_m is applied to a state it sets the value of the mth memory location to zero.

$$M_m \mid \ldots, n, \ldots > \equiv \frac{(a_m)^n}{\sqrt{n!}} \mid \ldots, n, \ldots >$$

m^{th} memory location value

$$= \quad | \ldots, 0, \ldots >$$

Note: The repeated application of factors of a_m to the state results in numerical factors that the denominator $\sqrt{n!}$ cancels.

Change Value at Memory Location m from 0 to Value at Location n

$$P_m{}^n = \frac{(a_m{}^\dagger)^{N_n}}{\sqrt{N_n!}}$$

The above expression for $P_m{}^n$ is also symbolic. The expression represents the following ordered operator expression in which the operators are carefully ordered.

$$P_m{}^n \equiv \sum_q \frac{(\ln a_m{}^\dagger)^q \, N_n{}^q}{q!} \frac{1}{\sqrt{N_n!}}$$

where the sum ranges from 0 to ∞. When $P_m{}^n$ is applied to a state it changes the value of the mth memory location from zero to the value at the nth memory location.

$$P_m{}^n \, | \ldots, 0, \ldots, x, \ldots > \equiv \frac{(a_m{}^\dagger)^x}{\sqrt{x!}} \, | \ldots, 0, \ldots, x, \ldots >$$

with labels m^{th} and n^{th} pointing to the positions.

$$= \quad | \ldots, x, \ldots, x, \ldots >$$

The application of $N_n{}^q$ to the state results in $\sqrt{x!}$ when the summation in $P_m{}^n$ is performed.

The operators M_m and $P_m{}^n$ enable us to simply express the algebraic equivalent of assembly language instructions:

LOAD m – load the value at memory location m into the register

$$P_0{}^m \, M_0$$

STORE m – store the value in the register at memory location m

$$P_m{}^0 M_m$$

SHIFT k – shift the value in the register by k bits. If k is positive the bit shift is to the right and if k is negative the bit shift is to the left. The bits are numbered from the leftmost bit which is bit 0 corresponding to 2^0. The next bit is bit 1 corresponding to 2^1 and so on.

If the bit shift is to the right (k > 0) then we assume the padding bits are 0's. For example a shift of the bit pattern for 7 = 1110000 ... by one bit to the right is 14 = 01110000 ... As a result the value in the register can be doubled (k = 1), quadrupled (k = 2), and so on. The algebraic expression for a k bit right shift is

$$\sum_q \frac{(\ln a_0{}^\dagger)^q S_0{}^q}{q!} \frac{\sqrt{N_0!}}{\sqrt{T_0!}}$$

where

$$S_0 = (2^k - 1)N_0$$

and

$$T_0 = 2^k N_0$$

If the bit shift is to the left (negative k), then we assume zero bits are added "at ∞". If k = -1 then the effect of left shift is to divide the value in the register by two (dropping the fractional part). If k = -2 then the effect of left shift is to divide the value in the register by four (dropping the fractional part) and so on. The algebraic expression that implements left shift is

$$\sum_q \frac{(\ln a_0)^q U_0{}^q}{q!} \frac{\sqrt{N_0!}}{\sqrt{V_0!}}$$

where

$$U_0 = N_0 - [2^k N_0]$$

and

$$V_0 = [2^k N_0]$$

97

with [z] being the value of z truncated to an integer (fractional part dropped).

ADD m – add the value at memory location m to the value in the register

$$\sum_q \frac{(\ln a_0{}^\dagger)^q \, N_m{}^q}{q!} \frac{\sqrt{N_0!}}{\sqrt{(N_0 + N_m)!}}$$

SUBTRACT m – subtract the value at memory location m from the value in the register (This expression assumes the value in the register is greater than or equal to the value at location m.)

$$\sum_q \frac{(\ln a_0)^q \, N_m{}^q}{q!} \frac{\sqrt{N_0!}}{\sqrt{(N_0 - N_m)!}}$$

MULTIPLY m – multiply the value in the register by the value at memory location m

$$\sum_q \frac{(\ln a_0{}^\dagger)^q \, (N_0)^q (N_m - 1)^q \, \sqrt{N_0!}}{q!} \frac{}{\sqrt{(N_0 N_m)!}}$$

DIVIDE m – divide the value in the register by the value at memory location m

$$\sum_q \frac{(\ln a_0{}^\dagger)^q \, W^q}{q!} \frac{\sqrt{N_0!}}{\sqrt{X!}}$$

where

$$W = N_0 - [\, N_0/N_m \,]$$

and

$$X = [\, N_0/N_m \,]$$

with [z] being the value of z truncated to an integer (fractional part dropped).

AND m – change the value in the register by "and-ing" it with the value at memory location m:

$$\sum_q \frac{[(\ln a_0^\dagger)^q (W)^q \theta(W) + (\ln a_0)^q (-W)^q \theta(-W)]}{q!} \frac{\sqrt{N_0!}}{\sqrt{X!}}$$

where

$$W = N_0 \& N_m - N_0$$

and where

$$X = N_0 \& N_m$$

with $\theta(z) = 1$ if $z > 0$ and 0 if $z < 0$. The & operator (adopted from the C programming language) performs bitwise AND. Corresponding bits in each operand are "multiplied" together using the bit multiplication rules:

$$1 \& 1 = 1$$
$$1 \& 0 = 0 \& 1 = 0 \& 0 = 0$$

For example the binary numbers 1010 & 1100 = 1000 or in base 10, 5&3 = 1.

OR m – change the value in the register by "or-ing" it with the value at memory location m:

$$\sum_q \frac{(\ln a_0^\dagger)^q (W)^q}{q!} \frac{\sqrt{N_0!}}{\sqrt{X!}}$$

where

$$W = N_0 | N_m - N_0$$

and where

$$X = N_0 | N_m$$

The | operator (adopted from the C programming language) performs bitwise OR. Corresponding bits in each operand are "multiplied" together using the multiplication rules:

$$1\,|\,1 = 1\,|\,0 = 0\,|\,1 = 1$$

$$0\,|\,0 = 0$$

For example the binary numbers $1010\,|\,1100 = 1110$ or in base 10, $5\,|\,3 = 7$.

NOT – change the value in the register by "not-ing" it:

$$\sum_{q} \frac{[(\ln a_0^{\dagger})^q (W)^q \theta\,(W) + (\ln a_0)^q\,(-W)^q \theta(-W)]}{q!} \quad \frac{\sqrt{N_0!}}{\sqrt{X!}}$$

where

$$W = \;\sim N_0 - N_0$$

and where

$$X = \;\sim N_0$$

with $\theta(z) = 1$ if $z > 0$ and 0 if $z < 0$.

The $\sim$ operator (adopted from the C programming language) performs bitwise NOT. Each 1 bit is replaced by a 0 bit and each 0 bit is replaced by a 1 bit. Since we have infinite words in our computer we supplement this rule with the restriction that the exchange of 1's and 0's only is made up to, and including, the rightmost 1 bit in the operand. The 0 bits beyond that remain 0 bits. For example the binary number $\sim 101 = 010$ or in base 10, $\sim 3 = 2$.

INPUT m – input a value storing it at memory location m. The input device is usually associated with a memory location from which the input symbolically takes place. We will designate the memory location of the input device as *in*.

$$P_m^{\;in}\,M_m$$

OUTPUT m – output the value at memory location m. The output device is usually associated with a memory location to which output symbolically takes place. We will designate the memory location of the output device as *out*.

$$P_{out}{}^m M_{out}$$

TRA m – set the value of the program counter to the value at memory location m. If we designate the program counter memory location as pc then this instruction is mapped to

$$P_{pc}{}^m M_{pc}$$

TZR m – set the value of the program counter to the value at memory location m if the value in the register is zero:

$$(P_{pc}{}^m M_{pc})^{\theta(N_r)\,\theta(-N_r)}$$

using $\theta(0) = 1$.

HALT – stop execution of the program. The halt in a program is mapped to a "bra" state vector.

$$< \ldots \,|$$

12.5 A Simple Assembly Language Program

Assembly language instructions can be combined to form an assembly language program. Perhaps the best way to see how the algebraic representation of assembly language works is to translate a simple assembly language program into its algebraic equivalent.

The program that we will consider is:

1   INPUT x
2   INPUT y
3   LOAD x
4   ADD y
5   STORE z
6   OUTPUT z
7   HALT

This program translates to the algebraic equivalent:

Steps: 7   6   5   4   3   2   1

$$< \dots \mid P_{out}{}^{z} M_{out} \;\; P_{z}{}^{0} M_{z} \;\; \frac{(a_{0}{}^{\dagger})^{N_{y}} \sqrt{N_{0}!}}{\sqrt{(N_{0} + N_{y})!}} \;\; P_{0}{}^{x} M_{0} \;\; P_{y}{}^{in} M_{y} \;\; P_{x}{}^{in} M_{x} \mid \dots >$$

where the power of $a_{0}{}^{\dagger}$ is represented by a power series expansion in $\ln(a_{0}{}^{\dagger})$ as seen earlier. Note the algebraic steps take place from right to left.

The algebraic expression in the brackets produces one output state from a given initial state. The values in memory after the last step correspond to one and only one output state of the form:

$$< n, m, p, \dots \mid \;\; = \;\; (N \, (r^{\dagger})^{n} (a_{0}{}^{\dagger})^{m} \, (a_{1}{}^{\dagger})^{p} \dots \mid 0 >)^{\dagger}$$

where N is a normalization constant.

This simple program does not produce a superposition of states. As a result programs of this type are analogous to ordinary programs for normal, non-Quantum computers. The numbers in memory after the program concludes are the "output" of the program.

We will see programs in succeeding sections that take a computer of "fixed" state $N(r^{\dagger})^{n}(a_{0}{}^{\dagger})^{m} \, (a_{1}{}^{\dagger})^{p} \dots \mid 0 >$ (or a superposition of states), and produce a superposition of states that must be interpreted quantum mechanically. These programs are quantum in nature and the computer that runs them must be a quantum computer.

12.6 Programs and Program Logic

The simple program of the last section corresponded to a sequential program that executed step by step. We now turn to more complex programs with program logic that supports non-sequential execution of programs. When this type of program executes some instructions can cause jumps from one instruction to another instruction in another part of the program.

Programs are linear – one instruction executes after another. But they are not necessarily sequential – the instructions do not always execute step by step sequentially. A program can specify jumps ("goto" instructions) in the code from the current instruction to an instruction several steps after the current instruction or several steps back to a previous instruction. The code then executes sequentially until the next jump is encountered.

These jumps in the code at the level of assembly language implement control constructs such as goto statements, if expressions, for loops, and switch expressions which are similar to control constructs in higher level languages such as C and C++.

Jumps in code can be implemented in the algebraic representation of programs by making a program's counter value jump in value. A step in a program will execute or not execute depending on the current value of the program counter.

Changes in the value of the program counter are made using the TRA and TZR instructions. In the algebraic representation the program counter variable can be used to manage the execution of the program steps.

The key algebraic constructs supporting non-sequential program execution are:

Execute instruction only if PC $\leq$ n

$$(\ldots)^{\theta \, (n \, - \, N_{pc})}$$

Execute instruction only if PC $\geq$ n

$$(\ldots)^{\theta \, (N_{pc} \, - \, n)}$$

Execute instruction only if PC = n

$$(\ldots)^{\theta\theta \, (N_{pc} \, - \, n)}$$

Execute instruction only if PC not equal to n

$$(\ldots)^{1 \, - \, \theta\theta(N_{pc} \, - \, n)}$$

where the parentheses contain one or more instructions, and where $\theta\theta(x) = 1$ if $x = 0$ and zero otherwise.

Using these constructs we can construct non-sequential programs that support "goto's", if's, and other control constructs.

To illustrate this feature of the algebraic representation we will consider an enhancement of the assembly language program seen earlier:

```
1       INPUT x
2       INPUT y
3       LOAD x
4       TZR y
5       ADD y
6       STORE z
7       OUTPUT z
8       HALT
```

This program has the new feature that if the first input, the input to memory location x is zero, then instruction 4 will cause a jump to the instruction specified by the value stored at memory location y.

For example if the inputs are 0 placed at memory location x and 2 placed at memory location y, then the TZR instruction (defined earlier) will cause the program to jump to instruction 2 from instruction 4. Then the program will proceed to execute from instruction 2.

Another example of a case with a jump is if the input to memory location x is zero and the input to memory location y is 6 then the program jumps from instruction 4 to instruction 6 and the program completes execution from there. If the input to memory location x is non-zero no jump takes place.

To establish the algebraic equivalent of the preceding example we have to use the non-sequential constructs provided earlier in this section. In addition we must define the algebraic equivalent *recursively* because of the possibility that the program may jump backwards to an earlier instruction in the program. If only "forward" jumps were allowed then recursion would not be needed.

An algebraic representation of a program that supports only forward leaps is:

$$< \dots \mid \overset{8}{} \overset{7}{(a_{pc}^{\dagger} P_{out}^{z} M_{out})}^{\theta\theta(Npc-7)}$$

$$\overset{6}{(a_{pc}^{\dagger} P_z^{r} M_z)}^{\theta\theta(Npc-6)}$$

$$\overset{5}{\left(\frac{a_{pc}^{\dagger} (a_r^{\dagger})^{N_y} \sqrt{N_r!}}{\sqrt{(N_r + N_y)!}} \right)}^{\theta\theta(Npc-5)}$$

$$\overset{4}{(a_{pc}^{\dagger})^{1-\theta\theta(Nr)} (P_{pc}^{y} M_{pc})^{\theta\theta(Nr)\, \theta\theta(Npc-4)}}$$

$$\overset{3}{a_{pc}^{\dagger} P_r^{x} M_r}$$

$$\overset{2}{a_{pc}^{\dagger} P_y^{in} M_y}$$

$$\overset{1}{a_{pc}^{\dagger} P_x^{in} M_x}$$

$$a_{pc}^{\dagger} M_{pc} \mid \dots >$$

where the superscript and subscript r represent the computer register. In step 4 the program counter value is set equal to the value of y. If y = 5 or 6 or 7 the program leaps forward to the statement of that number due to the $\theta\theta(N_{pc} - 5)$ etc. exponents of the subsequent statements. The program steps are numbered above each corresponding algebraic expression. The step function expressions enable a jump to take place successfully if the conditions are met.

A program, with both forward and backward jumps supported, requires a recursive definition. For example we can define the recursive function f():

$$\overset{7}{f() = (a_{pc}^{\dagger} P_{out}{}^{z} M_{out})}{}^{\theta\theta(Npc^{-} 7)}$$

$$\overset{6}{(a_{pc}^{\dagger} P_{z}{}^{r} M_{z})}{}^{\theta\theta(Npc^{-} 6)}$$

$$\overset{5}{(\frac{a_{pc}^{\dagger} (a_{r}^{\dagger})^{N_y} \sqrt{N_r!}}{\sqrt{(N_r + N_y)!}})}{}^{\theta\theta(Npc^{-} 5)}$$

$$\overset{4}{(a_{pc}^{\dagger})^{1-\theta\theta(Nr)} (f()P_{pc}{}^{y} M_{pc})}{}^{\theta\theta(Nr)\,\theta\theta(Npc^{-} 4)}$$

$$\overset{3}{(\,a_{pc}^{\dagger} P_{r}{}^{x} M_{r}\,)}{}^{\theta\theta(Npc^{-} 3)}$$

$$\overset{2}{(\,a_{pc}^{\dagger} P_{y}{}^{in} M_{y})}{}^{\theta\theta(Npc^{-} 2)}$$

$$\overset{1}{(\,a_{pc}^{\dagger} P_{x}{}^{in} M_{x})}{}^{\theta\theta(Npc^{-} 1)}$$

with the order of execution of the steps indicated by the integer above each factor. (Steps execute from right to left horizontally and/or bottom to top in the listing above.) A program using f() is

$$f()a_{pc}^{\dagger} M_{pc} \mid \ldots >$$

This program is well behaved except if the input value placed at the y memory location is 4. In this case the program recursively executes forever. This defect can be removed by using another memory location for a counter variable that limits the number of recursions.

We can modify the program so that the program only recursively calls itself a finite number of times by having each recursive call decrease a recursion counter variable by one. When the variable's value reaches zero the recursion terminates. An example of such a program (set to allow at most 10 iterations of the recursion) is:

$$\overset{7}{(a_{pc}{}^{\dagger}P_{out}{}^{z}\,M_{out})}{}^{\theta\theta(Npc^{-}7)}$$

$$\overset{6}{(a_{pc}{}^{\dagger}P_{z}{}^{r}\,M_{z})}{}^{\theta\theta(Npc^{-}6)}$$

$$\frac{\overset{5}{(a_{pc}{}^{\dagger}\,(a_{r}{}^{\dagger})^{N_y}\,\sqrt{N_r!}\,)}{}^{\theta\theta(Npc^{-}5)}}{\sqrt{(N_r + N_y)!}}$$

$$(a_{pc}{}^{\dagger})^{1-\theta\theta(Nr)}\,\overset{4}{((a_{pc}{}^{\dagger})^{1-\theta(Nw)}(g()a_w)^{\theta(Nw)}\,P_{pc}{}^{y}\,M_{pc})}{}^{\theta\theta(Nr)\,\theta\theta(Npc^{-}4)}$$

$$\overset{3}{(\,a_{pc}{}^{\dagger}P_{r}{}^{x}M_{r}\,)}{}^{\theta\theta(Npc^{-}3)}$$

$$\overset{2}{(\,a_{pc}{}^{\dagger}P_{y}{}^{in}\,M_{y})}{}^{\theta\theta(Npc^{-}2)}$$

$$\overset{1}{(\,a_{pc}{}^{\dagger}P_{x}{}^{in}\,M_{x})}{}^{\theta\theta(Npc^{-}1)}$$

The program is

$$g()a_{pc}{}^{\dagger}M_{pc}\,(a_w{}^{\dagger})^9 M_w \mid \ldots >$$

where w is some memory location. The factor of $(a_w{}^{\dagger})^9$ and the form of step 4 cause the iteration to end before or after 10 repetitions since the a_w factor within the inner parentheses of step 4 causes $\theta(N_W) = 0$ after 10 iterations. If the input y value is set to a value < 3 in the 10th iteration it then ends at the end of step 4. If the input y value is set to a value ≥ 3 in the 10th iteration it then proceeds to the step corresponding to the program counter value at the end of step 4.

Any assembly language program using the previously specified instructions can be mapped to an algebraic representation – possibly with the use of additional memory for values such as the w memory location value seen above.

Using the algebraic constructs supporting non-sequential program execution we can create algebraic representations of assembly language programs. These programs have a definite input state and through the execution of the program they evolve into a definite output state -–not a superposition of output states. Therefore they faithfully represent assembly language programs. On the other hand they are quantum in the sense that they use states and boson raising and lowering operators. The types of programs we are creating in this approach are "sharp" on the space of states. One input state evolves through the program's execution to one and only one output state with probability one.

These types of programs are analogous to free field theory in which incoming particles evolve without interaction to an output state containing the same particles.

In the next section we extend the ideas in this section to quantum programming where a variety of output states are possible – each with a certain probability of being produced.

12.7 Quantum Assembly Language™ Programs

In this section we will first look at a simplified quantum program that illustrates quantum effects but in actuality is a sum of deterministic assembly language programs mapped to algebraic equivalents. Consider a "quantum" program that is the sum of three ordinary programs $g_1()$, $g_2()$ and $g_3()$ of the type seen in the previous section. Further let us assume a set of orthonormal states

$$| \; n, m, p, \ldots >$$

similar to those that we saw in the previous sections with

$$< X \; | \; Y > = \delta_{XY}$$

where δ_{XY} represents a product of Kronecker δ functions in the individual values in memory of the $| \; X >$ and $| \; Y >$ states. Further let us assume

$$| \; n_1, m_1, p_1, \ldots > \; = g_1() \; | \; \ldots >$$

$$| \; n_2, m_2, p_2, \ldots > \; = g_2() \; | \; \ldots >$$

$$| \, n_3, m_3, p_3, \ldots > \, = g_3() \, | \, \ldots >$$

for some initial state $| \ldots >$ of the quantum computer. The initial state can be a pure state – a state of the form $| \, n, m, p, \ldots >$ in which each memory location has a specific value. Pure states are assumed to be suitably normalized to one: $< n, m, p, \ldots | \, n, m, p, \ldots > = 1$. Or it can be a superposition of pure states:

$$| \text{initial\_state}> \, = \sum_{n, m, p, \ldots} a_{n, m, p, \ldots} | \, n, m, p, \ldots >$$

where $a_{n, m, p, \ldots}$ are constants. The initial state and the set of possible final states are assumed to be normalized to one $< \ldots | \ldots > = 1$. Then

$$g() = a g_1() + \beta g_2() + \gamma g_3()$$

is a "quantum" program operator that takes an initial state of a computer into some final state, denoted $|f>$, where a, β, and γ are complex (in general), constant probability amplitudes satisfying

$$|a|^2 + |\beta|^2 + |\gamma|^2 = 1$$

We call

$$|f> \, = g() \, | \, \ldots >$$

a quantum computer program. The quantum program produces the state $| \, n_1, m_1, p_1, \ldots >$ with probability $|a|^2$, the state $| \, n_2, m_2, p_2, \ldots >$ with probability $|\beta|^2$, and the state $| \, n_3, m_3, p_3, \ldots >$ with probability $|\gamma|^2$ if the states are suitably normalized.

$$a = \, < n_1, m_1, p_1, \ldots | g() | \ldots >$$

$$\beta = \, < n_2, m_2, p_2, \ldots | g() | \ldots >$$

$$\gamma = \, < n_3, m_3, p_3, \ldots | g() | \ldots >$$

We now have a quantum probabilistic computer. The programs $g_1()$, $g_2()$ and $g_3()$ are being executed in *parallel* in a quantum probabilistic manner.

12.7.1 Classical Probabilistic Computer Programs

It is also possible to define a classical probabilistic computer and classical probabilistic computer program. The essence of the difference is using probabilities directly rather than probability amplitudes. An example of a classical probabilistic program operator is:

$$g_C() = a_C g_1() + \beta_C g_2() + \gamma_C g_3()$$

It takes an initial state of a computer $|i\rangle$ into some final state, denoted $|f\rangle$, where a_C, β_C, and γ_C are constant probabilities satisfying

$$a_C + \beta_C + \gamma_C = 1$$

We call

$$|f\rangle = g_C() |i\rangle$$

where $|i\rangle$ and $|f\rangle$ are pure states, or superpositions of pure states, the execution of a classical probabilistic computer program.

12.7.2 Creating Quantum Computers

Currently, the most feasible way of creating a Quantum Computer with current technology, or reasonable extrapolations of current technology, is to create a material which approximates a lattice with spins at each lattice site that we can orient electromagnetically at the beginning of a program. The execution of a program takes place by applying electromagnetic fields that have a time dependence specific to the computation. The electromagnetic fields implement a custom-tailored set of interactions between the spins in the material that simulates the calculation to be performed.

The interactions are specified with some Hamiltonian, or some effective Hamiltonian, and the initial state of the lattice spins evolves dynamically to some configuration that is then measured.

The Hamiltonians are normally specified in a space-time formalism that is a familiar part of Quantum Mechanics. A Hamiltonian specifies the time evolution of a system starting from an initial state. We can introduce an explicit time dependence in states by using the notation:

$$|\Psi(t)\rangle$$

to denote the state of a Quantum Computer at time t. The general state of the computer at time t can be written as a superposition of Fock representation states:

$$| \Psi(t) > = \sum_n f_n(t) | n_1, n_2, n_3, \ldots >$$

where n represents a set of values $n_1, n_2, n_3, \ldots$

12.7.3 Quantum Computer Programs and Hamiltonian Evolution Operators

The time evolution of the states can be specified using an exponentiated Hamiltonian operator H:

$$| \Psi(t) > = S | \Psi(0) > = e^{-iHt} | \Psi(0) >$$

With this Hamiltonian formulation we can simulate a physical (or mathematical) process by defining a Hamiltonian that corresponds to the process, and then creating an experimental setup using a set of lattice spins in some material that implements the simulation. The experimental setup will prepare the initial state of the spins, create a fine-tuned interaction that simulates the physics of the process, and then, after the system has evolved, will measure the state of the system at time t. Repeated performance of this procedure will determine the probability distribution associated with the final state of the Quantum Computer. The probability distribution is specified by the set of coefficients $| f_n(t) |^2$.

A simple example of a Hamiltonian that causes a Quantum Computer to evolve in a non-trivial way is:

$$H = \sum_{m=0}^{\infty} a_{m+1}^{\dagger} a_m$$

(This example was chosen partly because it has a form similar to a Virasoro algebra generator in SuperString Theory.) Let us assume the initial state of the Quantum Computer at $t = 0$ is

$$| 1, 0, 0, 0, \ldots >$$

that is, an initial value of 1 in the first bit in memory and zeroes in all other memory locations. At time t the state of memory is:

$$| \Psi(t) > = \sum_{n=0}^{\infty} f_n(t) | 0, 0, \ldots, \underset{\underset{\text{nth memory location}}{\downarrow}}{1}, 0, \ldots >$$

with

$$f_n(t) = (-it)^n/n!$$

using the power series expansion of the exponentiated Hamiltonian expression. The probability of finding the state

$$n^{th} \text{ memory location}$$

$$\downarrow$$

$$| 0, 0, \ldots, 1, 0, \ldots >$$

is

$$(t^n/n!)^2$$

At first glance the Hamiltonian approach is very different from the Quantum Assembly Language™ approach discussed above. However these approaches can be interrelated in special cases and (we believe) in the general case through sufficiently clever transformations. For example, the preceding Hamiltonian can be re-expressed as assembly language instructions

$$H = \sum_{m=0}^{\infty} (\text{STORE } (m+1))(\text{ADD } 1)(\text{LOAD } (m+1)) \ (\text{STORE } m) \ (\text{SUBTRACT } 1)(\text{LOAD } m)$$

where a value is loaded into the register from memory location m and then 1 is added to the value in the register. The 1 appearing in the expressions "ADD 1" and "SUBTRACT 1" above represents a literal numeric value one – not a memory location. The parentheses around m+1 in LOAD(m + 1) indicates it is the $(m+1)^{th}$ memory location – not the addition of one to the value at the mth location.

The preceding assembly language expression for H can be replaced with the algebraic representation expression:

$$H = \sum_{m=0}^{\infty} P_{m+1}{}^r M_{m+1} \, a_r^\dagger \, \frac{1}{\sqrt{(N_r + 1)}} \ P_r{}^{m+1} M_r P_m{}^r M_m a_r \sqrt{N_r} P_r{}^m M_r$$

where r represents the computer's register. This complex expression is not an improvement in one sense. The original Hamiltonian expression was much simpler. Its importance is the mappings that it embodies from a quantum mechanical Hamiltonian to an assembly language expression to an algebraic representation of the assembly language.

If we regard the value in the register as a "scratchpad" value, as programmers often do, then we can establish a representation of $a_m{}^\dagger$ and a_m in terms of the algebraic representation of assembly language instructions.

$$a_m{}^\dagger \equiv P_m{}^r \, M_m \, a_r{}^\dagger \frac{1}{\sqrt{(N_r + 1)}} \, P_r{}^m \, M_r$$

and

$$a_m \equiv P_m{}^r \, M_m \, a_r \, \sqrt{N_r} \, P_r{}^m \, M_r$$

The power series expansion of the exponentiated Hamiltonian in the previous example is an example of the use of Perturbation Theory. The direct solution of a problem is often not feasible because of the complexity of the dynamics. Physicists have a very well developed theory for the approximate solution of these difficult problems called Perturbation Theory. Perturbation Theory takes an exact solution of a simplified version of the problem and then calculates corrections to that solution that approximate the exact solution of the problem.

In the preceding example the initial state of the Quantum Computer represents a time-independent description of the Quantum Computer. The time-dependent description of the Quantum Computer, which is the sought-for solution, requires the evaluation of the result of the application of the exponentiated Hamiltonian to the initial state. For a small elapsed time, the exponential can be expanded in a power series and the application of the first few terms of the power series to the initial state approximates the actual state of the Quantum Computer. Thus we have a Perturbation Theory for the time evolution of the Quantum Computer expressed as an expansion in powers of the elapsed time. A common approximation is to approximate the evolution operator S with

$$S \simeq 1 - iHt$$

with H the Hamiltonian in order to get a first order approximation to the quantum program.

12.8 Bit-Level Quantum Computer Language

In the previous section we examined a Quantum Assembly Language™ with words consisting of an infinite sets of bits. In this section we will examine the opposite extreme – a Quantum Computer Language with one-bit words. One can also create Quantum Computer Languages for intermediate cases such as 32-bit words.

A Bit-Level Quantum Computer Language can be represented with anti-commuting Fermi operators b_i and $b_i^\dagger$ for $i = 0, 1, 2, \ldots$ representing each bit location in the Quantum Computer's memory. The operators have the anti-commutation rules:

$$\{ b_i, b_j^\dagger \} = \delta_{ij}$$
$$\{ b_i, b_j \} = 0$$
$$\{ b_i^\dagger, b_j^\dagger \} = 0$$

where δ_{ij} is 1 if $i = j$ and zero otherwise. We will assume an (unrealistic) one-bit register with a pair of raising and lowering operators r and $r^\dagger$ for the register with the anti-commutation relations:

$$\{ r_i, r_j^\dagger \} = \delta_{ij}$$
$$\{ r_i, r_j \} = 0$$
$$\{ r_i^\dagger, r_j^\dagger \} = 0$$

The ground state of the computer is the state with values set to zero at all bit memory locations. It is represented by the vector

$$| 0, 0, 0, \ldots > \equiv | 0 > \equiv \Phi_V$$

A typical state of the computer can be represented with a vector such as

$$| 1, 1, 1, \ldots > = r^\dagger b_0^\dagger b_1^\dagger \ldots | 0 >$$

with the first number being the value in the register, the second number the value at memory location 0, the third number the value at memory location 1, and so on.

A specified Quantum Computer state evolves as a Quantum Computer Program executes to a final computer state. A Bit-Level Quantum Computer Program can be represented as an algebraic expression in anti-commuting raising and lowering operators. The approach is similar to the approach seen earlier in this chapter for infinite-bit words using commuting operators.

12.9 Basic Operators of the Bit-Level Quantum Language

The key operators that are required for the algebraic representation of a Bit-Level Quantum Computer Language™ are:

<u>Fetch the Value at a Memory Location (Number Operator)</u>

$$N_m = b_m{}^\dagger b_m$$

For example,

$$N_m \mid \ldots, 1, \ldots > = \mid \ldots, 1, \ldots >$$

m^{th} memory location value

<u>Set the Value at Memory Location m to Zero</u>

$$M_m = (b_m)^{N_m}$$

The above expression for M_m is symbolic. The expression represents the following expression in which the operators are carefully ordered to avoid complications (c-numbers etc.) resulting from reordering.

$$M_m \equiv e^{N_m \ln b_m} = \sum_q \frac{(\ln b_m)^q N_m{}^q}{q!}$$

where the sum ranges from 0 to ∞. M_m becomes

$$M_m = 1 + (b_m - 1)N_m$$

using the identity $N_m = N_m{}^2$. When M_m is applied to a state it sets the value of the mth memory location to zero.

$$M_m \mid \ldots, x, \ldots > = \mid \ldots, 0, \ldots >$$

m^{th} memory location value

<u>Change Value at Memory Location m from 0 to Value at Location n</u>

$$P_m{}^n = (b_m{}^\dagger)^{N_n}$$

The above expression for $P_m{}^n$ is also symbolic. The expression represents the following expression in which the operators are carefully ordered to avoid complications (c-numbers etc.) resulting from reordering.

$$P_m{}^n \equiv \sum_q \frac{(\ln b_m{}^\dagger)^q \, N_n{}^q}{q!}$$

where the sum over q ranges from o to ∞. Using the identity $N_m = N_m{}^2$ the expression for $P_m{}^n$ simplifies to:

$$P_m{}^n = 1 + (b_m - 1)N_m$$

When $P_m{}^n$ is applied to a state it changes the value of the mth memory location from zero to the value at the nth memory location.

$$P_m{}^n \,|\, \ldots, \overset{m^{th}}{0}, \ldots, \overset{n^{th}}{x}, \ldots > = \quad (b_m{}^\dagger)^x \,|\, \ldots, 0, \ldots, x, \ldots>$$

$$= \quad |\, \ldots, x, \ldots, x, \ldots >$$

We can use the operators M_m and $P_m{}^n$ to express bit-wise assembly language instructions:

LOAD m – load the value at memory location m into the register

$$P_r{}^m \, M_r = (1 - N_r + b_r)(1 - N_m) + (N_r + b_r{}^\dagger)N_m$$

The first term on the right handles the case $N_m = 0$ and the second term on the right handles the case $N_m = 1$.

STORE m – store the value in the register at memory location m

$$P_m{}^r \, M_m = (1 - N_m + b_m)(1 - N_r) + (N_m + b_m{}^\dagger)N_r$$

The first term on the right handles the case $N_r = 0$ and the second term on the right handles the case $N_r = 1$.

115

ADD m – add the value at memory location m to the value in the register

$$(b_r^\dagger)^{Nm} = \sum_q \frac{(\ln b_r^\dagger)^q \, N_m{}^q}{q!}$$

$$= 1 + (b_r^\dagger - 1)N_m$$

If both the register and memory bit m have values of one then the application of this operator expression to the quantum state produces zero.

SUBTRACT m – subtract the value at memory location m from the value in the register

$$(b_r)^{Nm} = \sum_q \frac{(\ln b_r)^q \, N_m{}^q}{q!}$$

$$= 1 + (b_r - 1)N_m$$

If the value in the register is zero and the value at location m is one the application of this operator produces zero.

MULTIPLY m – multiply the value in the register by the value at memory location m

$$(b_r^\dagger)^{(Nm-1)Nr} = \sum_q \frac{(\ln b_r^\dagger)^q \, (N_r)^q (N_m - 1)^q}{q!}$$

$$= 1 + (b_r - N_r)(1 - N_m)$$

Other assembly language instructions can be expressed in algebraic form as well.

The operator algebra that we have developed for Quantum Assembly Language™ and Quantum Machine Language™ provides a framework for the investigation of the properties of Quantum Languages within an algebraic framework – a far simpler task than the standard quantum linguistic approaches.

12.10 Quantum High Level Computer Language Programs

Our Quantum Assembly Language™ representation forms a basis for the development of high level Quantum Programming Languages. These languages are quantum analogs of high-level computer languages such as C or C++ or FORTRAN.

In ordinary computation a statement in a high level language such as

$$a = b + c;$$

in C programming is mapped to a set of assembly language by a C compiler. A simple mapping of the above C statement to assembly language would be

LOAD ab
ADD ac
STORE aa

where aa is the memory address of a, ab is the memory address of b and ac is the memory address of c.

If we decide to define a High Level Quantum Computer Language™ then it would be natural to define it analogously in terms of a Quantum Assembly Language™. A statement in the High Level Quantum Computer Language™ would map to a set of Quantum Assembly Language™ instructions.

For example, a = b + c would map to the algebraic expression

$$P_{aa}{}^r M_{aa} \ \left(a_r{}^\dagger\right)^{N_{ac}} \frac{\sqrt{N_r!}}{\sqrt{(N_r + N_{ac})!}} \ P_r{}^{ab} M_r$$

using the formalism developed earlier in this chapter to LOAD, ADD and STORE.

The definition of high level Quantum Computer Languages™ in this approach is straightforward. One can then imagine creating programs in these languages for execution on Quantum Computers just as ordinary programs are created for ordinary computers.

Another approach to higher level Quantum Computer Languages™ is to simply express them directly using raising and lowering operators – not in terms of Quantum Assembly Language™ instructions. For example the preceding a = b + c; statement can be directly expressed as

$$\left(a_{aa}{}^\dagger\right)^{N_{ac}+N_{ab}} \left(a_{aa}\right)^{N_{aa}} \frac{1}{\sqrt{(N_{ac} + N_{ab})!} \ \sqrt{N_{aa}!}}$$

where aa is the memory address of a, ab is the memory address of b and ac is the memory address of c.

Simple High Level Quantum Computer programs can be expressed as products of algebraic expressions embodying the statements of the program. These programs are sharp on the set of memory states taking an initial memory state that is an eigenstate of the set of number operators N_m into an output eigenstate of the number operators.

The more general form of High Level *Quantum* Computer Program is a sum of deterministic High Level Programs (programs that take each input pure state into an output pure state). These programs would be similar in form to the Quantum Assembly Language case considered earlier. For example,

$$h(\,)|\,\ldots> \,=\, (ah_1(\,) + \beta h_2(\,) + \gamma h_3(\,))|\,\ldots>$$

where a, β, and γ are constants such that

$$|a|^2 + |\beta|^2 + |\gamma|^2 = 1$$

Each term $h_i(\,)$ transform an input pure state (eigenstate of all the number operators N_m) into an output pure state.

An initial eigenstate of the number operators is transformed into an output state that is a superposition of number operator eigenstates. A probability amplitude is associated with each possible output number operator eigenstate. The square of the amplitude – the probability – is the likelihood that the output eigenstate will be found when the output state of the Quantum Computer is measured.

A Hamiltonian can also be used to specify the time evolution of a system starting from an initial state just as in the Quantum Assembly Language case. Using the notation:

$$|\Psi(t)>$$

to denote the state of a Quantum Computer at time t the general state of a quantum computer at time t can be written as a superposition of number representation states:

$$|\Psi(t)> \,=\, \Sigma_n f_n(t)|\,n_1, n_2, n_3, \ldots>$$

where n represents a set of values $n_1, n_2, n_3, \ldots$

The time evolution of the states can be specified using the exponentiated Hamiltonian operator H:

$$| \Psi(t) > = e^{-iHt} | \Psi(0) >$$

A simple example of a Hamiltonian that causes a Quantum Computer to evolve in a non-trivial way is:

$$H = \sum_{m=0}^{\infty} (a_{m+2}^{\dagger})^{N_{m+1}+N_m} (a_{m+2})^{N_{m+2}} \frac{1}{\sqrt{(N_{m+1}+N_m)!} \sqrt{N_{m+2}!}}$$

This Hamiltonian performs additions similar to the a = b + c statement above. It generates a complex superposition of states as time evolves. More complex Hamiltonians equivalent to programs with several statements can be easily constructed.

12.11 Quantum C Language

One of the most important computer languages is the C programming language developed at Bell Laboratories in the 1970's. The original version of version of the C language was a remarkable combination of low level (assembly language-like) features and high level features like the mathematical parts of FORTRAN. The variables in the language were integers stored in words just as we saw in the earlier examples in this chapter. (There were several other types of integers as well – a complication that we will ignore in the present discussion.)

The C++, Java, and C# languages evolved from the C language. They preserve many C language features.

Using the ideas seen in the earlier sections of this chapter it is easy to develop algebraic equivalents for most of the constructs of the C language and thus create a Quantum C Language™. An important element that must be added to the previous development is to introduce the equivalent of pointers. Simply put pointers are variables that have the addresses of memory locations as their values. The C language has two important operators for pointer manipulations:

Operator	Role	Example
&	Fetch an address	ptr = &x;
*	Fetch the value at an address	z = *ptr;
	Set the value at an address	*ptr = 99;

The & operator of C fetches the address of a variable in memory. The example above shows a pointer variable ptr being set equal to the address of the x variable.

The * (dereferencing) operator can fetch the value at a memory location. The first * example above illustrates this aspect: the variable z is set equal to the value at the memory location specified by the pointer variable ptr.

The * operator can also be used to set the value at a memory location as illustrated by the second * example above. In this example the value 99 is placed at the memory location (address) specified by the ptr pointer variable.

These operators can be implemented in the algebraic representation of the Quantum C Language™ in the following way:

$$\& \quad \Longleftrightarrow \quad [A,\ ]$$

where $A = \Sigma\, m(a_m - a_m^\dagger)$ with the sum from 0 to ∞, and where the [A,] notation signifies forming a commutator. If we apply the [A,] operator to a raising or lowering operator we obtain its address:

$$\& a_m^\dagger = [A, a_m^\dagger] = m = \& a_m = [A, a_m]$$

The equivalent of the * operator is actually a pair of operator expressions. To fetch the value at a memory location m we use

$$*m \equiv N_m$$

The number operator N_m fetches the value at memory location m. Therefore the C code

$$y = *ptr;$$

where ptr is an integer address in memory, becomes the algebraic expression

$$(a_{ay}^\dagger)^{N_{ptr}}$$

where ay is the memory address of the y variable and N_{ptr} is the value at memory location ptr (which is an integer memory address).

To set the value at memory location m to the value X in C we write the code:

$$*m = X;$$

The equivalent algebraic expression is:

$$*m(X) \equiv (a_m^{\dagger})^X \, (a_m)^{N_m} \, \frac{1}{\sqrt{X!} \, \sqrt{N_m!}}$$

where *m(X) is a functional notation for the algebraic expression on the right side. In words, *m(X) means set the value at the address specified by m to the value X.

An (admittedly artificial) example of the use of these operators is the "pointer" algebraic expression for the C code statement:

$$a = b + c;$$

The corresponding "pointer" algebraic expression is:

$$\left(a_{aa}^{\dagger}\right)^{*ac \, + \, *ab} \left(a_{aa}\right)^{*aa} \frac{1}{\sqrt{(*ac + *ab)!} \, \sqrt{*aa!}}$$

where aa is the address of a, ab is the address of b, and ac is the address of c, and where *aa is the value at the address of a, *ab is the value at the address of b, and *ac is the value at the address of c. This expression can also be written as

$$\left(a_{[A, \, a_a]}^{\dagger}\right)^{N_{ac} \, + \, N_{ab}} \left(a_{[A, \, a_a]}\right)^{N_{aa}} \frac{1}{\sqrt{(N_{ac} + N_{ab})!} \, \sqrt{N_{[A, \, a_a]}!}}$$

Or more compactly using the functional notation as

$$*aa(*ab + *ac)$$

The Quantum C Language™ could be used to define Hamiltonians for a Quantum Computer. Other languages, such as Java™, C++, lisp and so on, also have Quantum analogues, which may be defined in a similar way.

12.12 SuperString Quantum Computer Revisited

The concepts we have been exploring for the algebraic representation of Quantum Programs raise the question - Can the SuperString Quantum Computer™, and Superstring interactions in particular, be cast into the form of algebraic quantum computations.

It appears that it may be possible in many SuperString theories to accomplish this purpose. The basis for this formulation is first the use of the raising and lowering operator formalism as the primary formalism and the relegation of the space-time formalism of SuperStrings to a secondary status. Secondly, the state-operator isomorphism of SuperString theory can be used to develop a representation of vertex operators for interactions as as states within the processor part of the SuperString Quantum Computer™. This surprising feature of SuperString theory answers a very basic question confronting the SuperString Quantum Computer™: How do we represent SuperString interactions in the Quantum Computer framework. The answer appears to be that we treat interactions (vertex operators) as intermediate operators within the Quantum Computer. Without this feature of SuperString theory the representation of interactions in the SuperString Quantum Computer™ would be an open question.

13. A Metatheory of Physics Theories

13.0 The State of Physics Theories

There are numerous theories in Physics. Most theories are specific to a certain domain of physical phenomena. For example there is the Standard Model that describes electromagnetic interaction, weak interaction and strong interaction phenomena. We will call theories that are specific to a certain set of physical phenomena, *phenomena domain theories*. Some examples of phenomena domain theories are the Standard Model, Newtonian Mechanics, General Relativity, superfluid He^3 theory, quantum mechanics, and BCS theory. Often, parts of several phenomena domain theories are required to explicate a set of physical phenomena. Consequently, phenomena domain theories can be combined to create a new, more general, phenomena domain theory.

There is another type of physics theory that probably originated in Greece, India or China. Today some physicists call it "The Final Theory" (Weinberg (1992)) or "The Theory of Everything." This type of theory purports to be a fundamental theory of all phenomena of nature. Every valid phenomena domain theory is a special case of it that is, in principle, ultimately derivable from the Theory of Everything.

We will use T_E to designate the Theory of Everything from this point on.

Currently there are two types of possible T_E's that are considered viable: Superstring theories, and extended embodiments of the Standard Model quantum field theory united with quantum gravity. It is possible that the true T_E, if found, may be of neither of these types but of a totally unforeseen type.

Progress with respect to Superstring theories, and quantum field theory-based theories, since the early 1970's has not been encouraging. While there have been major technical theoretical advances in both types of theories there are no decisive candidates for THE T_E in either type of theory. In particular, there is an embarrassing lack of experimental evidence for Superstring theories in general and for Standard Model extensions such as Technicolor theories. While the experimental situation may change in the next ten years or so, it is unlikely to lead to experimental results that would produce a decisive advance towards the goal of a T_E.

Faced with this impasse we will try to determine characteristics of THE T_E, should it exist. We could take a direct approach[17] and extrapolate current features of

---

[17] Weinberg(1992).

particle theories to get some idea of the T_E. This approach is reasonable and does lead to a short list of features that are likely to be present in the T_E. (See Weinberg (1992).)

However there are major questions about the T_E that cannot be answered in a direct, extrapolation-based approach. First among these questions is does a T_E exist? This question is relevant because mathematics – its close sister science – does not have an equivalent to a T_E as far as we presently know.[18] If mathematics, which is based on "pure" thought, does not have an all-embracing foundation, then can we expect physics, which has the added burden of experimental verification to have a T_E? And since we expect a T_E to be mathematical do not the foundational shortcomings of mathematics carry over into a physics T_E?

Clearly this fundamental question, as well as other fundamental questions such as consistency, requires investigation. In addition, we can hope that illumination might be shed on more practical issues such as why physics is fundamentally quantum (if it is).

To this end we will develop a metatheory of physics theories that is similar in concept to the metatheory of mathematics (*metamathematics*). A metatheory of physics has an added degree of complexity since it has associated experimental phenomena. Yet we can hope to uncover features and restrictions on the T_E.

13.1 The Metatheory of Mathematics (Metamathematics)

We will begin with a brief overview of the metatheory of mathematics. The reader should be warned that there are several different sets of terminology used in its development and several different schools of thought (the *formalists* or *axiomaticists* most of whom follow Hilbert's thinking, and the *intuitionists* whose guiding light was L. E. J. Brouwer).[19]

While there are different thoughts as to "a priori concepts", and the methods and procedures that constitute a metatheory, there is a commonality in the overall structure of the metatheory of mathematics.

This commonality is based on taking a mathematical-deductive system – which consists of a set of primitive terms, a set of axioms, a set of rules of inference, and a set of theorems – and abstracting the logical structure of the mathematical-deductive system to obtain a *formal system* (also called a *calculus*) involving only abstract symbols.

The calculus expresses the logic and theorems of the mathematical-deductive system as a series of formulas. These formulas start with the axioms and lead from them through intermediate formulas to the various theorems that, together with the axioms and rules of inference, comprise the calculus. Every theorem in the mathematical-deductive system corresponds to a formula (theorem) in the calculus in a one-to-one manner. Thus one can use the calculus to prove theorems of the

---

[18] p. 189ff Kleene (1967).
[19] See Kleene (1967) and Curry (1977) for more extensive discussions.

mathematical-deductive system. And the sequence of formulas that prove a theorem in the calculus is one-to-one mappable to the sequence of statements that prove the corresponding theorem in the mathematical-deductive system. Thus the calculus is like an "x-ray" of the mathematical-deductive system that shows its "skeleton" of logic and interconnectedness.

As we will see in the next chapter, Gödel used a calculus to great advantage to prove his celebrated Undecidability Theorem. In proving the theorem he developed a map from formulas to numbers called Gödel numbers that enabled the proof to be cleverly constructed.

Mathematical-Deductive System   **Calculus**

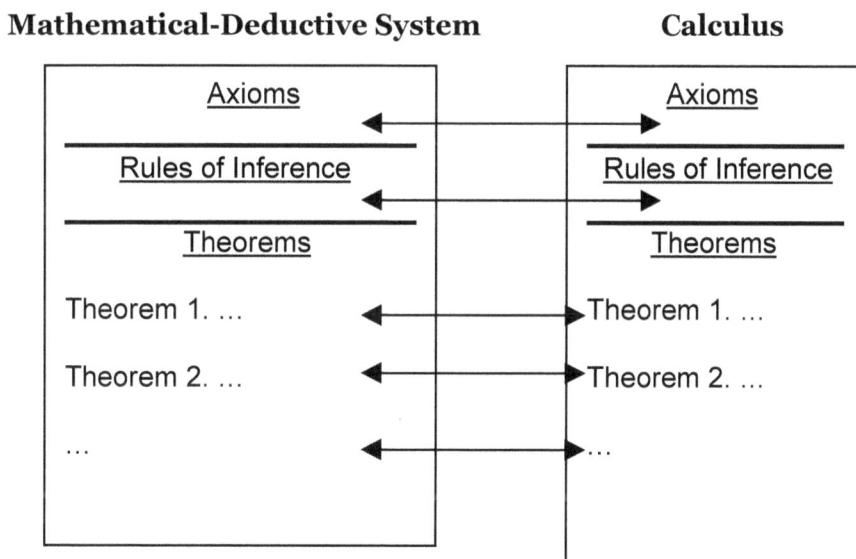

Figure 13.1.1. The one-to-one map of the Mathematical-Deductive System elements to formulas of the corresponding Calculus.

13.1.1 Mathematical-Deductive System

A mathematical-deductive system is, as indicated earlier, a set of primitive terms, a set of axioms, a set of rules of inference, and the set of all theorems provable from the axioms using the rules of inference. It is assumed that there are a finite number of rules of inference (and of axioms).

The primitive terms can be specified or left as undetermined items. An example of the second case is group theory in which the set of elements and the multiplication rules are not specified. The resulting set of theorems can be applied in numerous theoretic contexts.

An example of the first case is Euclidean geometry where primitives such as point, line and plane are specified giving meaning to the axioms so that questions such as the independence of the parallel postulate can be examined. A branch of logic known as *formal axiomatics* is based on studying a mathematical-deductive system as a "function" of the meaning of the primitive terms.

Other issues that confront a mathematical-deductive system are the consistency of the axioms and the decidability of all theorems that can be constructed from the primitives. Gödel's Undecidability Theorem shows that statements can be created using the primitives of some mathematical-deductive systems that cannot be proved or disproved. Chapter 14 discusses this point in some detail.

13.1.2 Formal System or Calculus

A formal system or calculus can be generated from a mathematical-deductive system by replacing each primitive term with a symbol so that each axiom becomes a symbolic formula. The rules of inference are also represented symbolically. The symbolic representation of the rules of inference enables a formula to be symbolically transformed into a new formula. Thus the proof of a theorem starts from an axiom or previously proven theorem, and through the repeated "mechanical" (Gödel's term) application of the symbolic rules of inference changes the initial formula to a new formula. It proceeds from formula to formula until a formula is reached that is the theorem to be proven.

The calculus also includes all other possible well-formed formulas that can be formulated in terms of the primitive and defined terms of the calculus. *A well-formed formula is a formula that is consistent with all the syntactical rules of the calculus.* Thus the equation x + + 3 = 5 is not a well-formed formula while x + 3 = 5 is a well-formed formula.

Therefore we can view a calculus as generated according to the following one-to-one map plus any additional well-formed formulas:

Mathematical-Deductive System		Formal System - Calculus
Primitive Term	→	Symbol
Axiom	→	Axiom formula
Rule of Inference	→	Symbolic Rule of Inference
Theorem	→	Formula
Theorem Proof step	→	Formula step
Consecutive Proof steps	→	Consecutive formulas related by a Symbolic Rule of Inference
Theorem Proof	→	A series of formulas ending in the formula equivalent to the desired Theorem
		All other possible well-formed formulas

It is also possible to consider a calculus in its own right – independent of any Mathematical-Deductive System and develop the set of theorems generated from the axiom formulas. This approach will not be relevant for the metatheory of physics theories.

13.2 Framework of a Metatheory of Physics

Physics theories are developed through a process of scientific induction with a constant interplay between theory and experiment. The hallmark of a *mature physics theory*, since the time of Newton is a mathematical-deductive structure. By that, I mean that a mature theory has a set of primitive terms and a set of axioms, or lagrangian, from which mathematical deductions are made to obtain experimental predictions. The number of primitive terms is usually not large and the number of axioms is also usually not large. The mathematical deductions are sometimes rigorous, sometimes not totally rigorous, and sometimes heuristic – meaning they make physical sense although they cannot be rigorously proved.

Perhaps the best example of a mature physical system is Newton's mechanics. It has a few primitive terms mass, acceleration, and so on. It has a few axioms such as F = ma. From these axioms the non-relativistic mechanics of point-like and extended materials bodies can be determined.

A little remarked fact of Newton's mechanics is that it was a heuristic theory for hundreds of years until the latter half of the 19th century when Dedekind and others established a rigorous foundation for the calculus of derivatives and integrals. Today we have a similar situation with respect to path integrals, which appear in many parts of physics including gauge theories and theories of quantum gravity. Path integral formulations work, but we do not have a rigorous mathematical foundation for them. [20]

Therefore we will assume mature physical theories are mathematical-deductive systems consisting of axioms and a set of statements/theorems/equations (abbreviated **STE**) that are all rigorously proved (although some proofs may currently only be proofs in the physicist's sense of the word, which includes heuristic proofs.) We will assume that rigorous proofs will exist in the future for all STE's of a correct mature theory.

Another issue that appears both in the metatheory of mathematics and physics is the issue of ideal vs. real statements. [21] A *real statement* is a statement that is finite in nature and therefore "real." In physics it is an experimentally testable statement. In mathematics it is a statement involving finite entities that has an intuitive meaning. An *ideal statement* is a statement that is an extension of a statement about finite entities to the case of the infinite. In physics an example of an ideal statement is the sum of the

---

[20] The metatheory of mathematics has a number of fundamental issues that should give comfort to physicists who worry about the lack of rigor in some physical theories, and should give pause for thought to those physicists who place too much weight on rigor. H. Curry on p. 7 of Curry (1976) points out, "The absolute rigor, which was then thought to have been obtained by the arithmetization of analysis, turns out to be a delusion." There are still fundamental questions about the nature of rigorous mathematical proof.
[21] p. 197 Kleene (1967).

probabilities of all possible outcomes of a quantum experiment is one *for all energies*. Since we cannot do an experiment at infinite energy the phrase "for all energies" makes the preceding statement an ideal statement. The statement, "The sum of the probabilities of all possible outcomes of a doable quantum experiment is one", is a real statement. Hilbert, and most physicists, would accept ideal statements as part of a mathematical-deductive system since such statements often simplify the theorems. *Therefore we will include ideal statements within the set of STE's of a mature physical theory*.

Before describing the axioms of the metatheory of physics we will briefly examine the various types of physics theories.

13.2.1 Deterministic Theories, Quantum Theories, Hybrid Theories

A deterministic theory is a theory in which all provable theorems (STE's) make predictions that can be experimentally tested in the phenomena of the theory (or are ideal statements), and in which all primitives and physical quantities are experimentally measurable "simultaneously" in principle. By simultaneously we mean within the course of an experiment with the measurement of one quantity *not necessarily* changing the value of a previously measured quantity. The classic example of a deterministic theory is Newton's mechanics.

A quantum theory is a theory that asserts that the measurement of some combinations of two quantities is incompatible in the sense that a measurement of one quantity invalidates the measurement of the other quantity. The simplest example of this type of theory is quantum mechanics in which the simultaneous measurement of a particle's position and momentum is not possible. Quantum theories have observable parts (experimentally measurable) and unobservable parts (not experimentally measurable). Thus we can divide the set of all STE's of a quantum theory into two parts as shown in Fig. 13.2.1.1.

A hybrid theory is a theory containing both quantum and non-quantum sectors. Phenomenological theories of this type are constructed to model the interaction of a quantum entity with a classical entity. A simple model theory of the photoelectric effect with incoming electrons causing photons (quantum) to be ejected from the surface of a metal (treated classically).

The division of physics theories into quantum, deterministic and hybrid raises the question of whether there are physics theories that are not of these three types. The answer is obviously no. Thus we assert the lemma:

Lemma 13.2.1.1. A physics theory is either a quantum theory, or a deterministic theory or a hybrid theory with quantum and deterministic sectors.

Proof: Consider an arbitrary physics theory. If there is no uncertainty relation (all operators commute if there are operators) then the theory is deterministic. If there is at least one uncertainty relation (pair of non-commuting operators) then the theory is either quantum or hybrid. Therefore it can consist of a quantum sector, a deterministic sector and possibly interactions between these sectors. If there is no purely deterministic sector then the theory is quantum. Otherwise it is hybrid. If the theory cannot be separated into independent sectors with possible interactions between them then it is quantum.

■

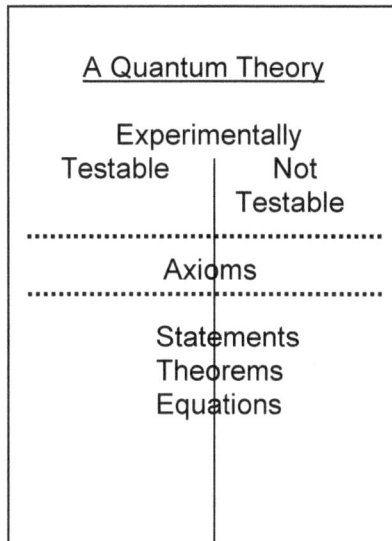

Figure 13.2.1.1. Some of the axioms and statements/theorems/equations (STE) are experimentally testable (observable) and some are not (unobservable).

13.2.2 Statistical Theories

There are physical theories that are statistical in nature such as the theory of ideal gases. These theories are based on an underlying dynamics that may be deterministic, quantum, or hybrid. We therefore classify statistical theories as quantum, deterministic or hybrid depending on their underlying dynamics.

13.2.3 Lagrangian Theories vs. non-Lagrangian Theories

Physics theories can be divided into two types: theories based on a lagrangian formulation, and theories that cannot be based on a lagrangian. There are many examples of lagrangian-based theories. Non-lagrangian theories can take the form of a set of axioms or a set of dynamical equations not susceptible to a lagrangian formulation.

The rules of inference in the case of lagrangian-based theories will include lagrangian-related procedures to obtain the dynamical equations of motion as well conservation laws and other items of physical interest.

The rules of inference for non-lagrangian theories will obviously be different from those of lagrangian-based theories. Their exact nature depends on the details of the theory. One requirement is the number of rules of inference is finite.

13.2.4 Rules of Inference for Physics Theories

The rules of inference for physics theories contain the equivalent of the rules of inference for the mathematical formalisms that are used in their construction. In addition, each theory can contain rules specific to the theory. Rules are allowed to have any mathematical, algebraic, derivative and integration constructs. The only restrictions are

1. The number of rules must be finite.
2. Each rule must be translatable to a formal system rule consisting of a finite number of symbols.

Rules can contain recursive functions and recursive constructs.

13.2.5 The Set of All Physics Theories

The set of all physics theories is defined to contain all physics theories that have a finite number of primitive terms, a finite number of symbols in their axioms and/or lagrangian, and a finite number of rules of inference. Recursively-defined axioms, and lagrangians containing recursively-defined functions, are allowed. All known physics theories meet these requirements. The set of all physical theories includes an *empty theory* that contains no primitives, axioms or theorems. (It has an *empty calculus*.)

13.3 Axioms of a Metatheory of Physics Theories

Our comments at the beginning of section 13.2 suggest that mature physical theories should be assumed to be rigorously provable mathematical-deductive systems for the sake of developing a metatheory of physics theories whether they are currently rigorous or will be at some time in the future. We will therefore assume the following axioms for a metatheory of physics theories:

(I)   Physical theories are mathematical-deductive systems consisting of a set of statements/equations that are derivable from a set of axioms that consist of a small set of postulates and/or a lagrangian. The axioms are based on a set of primitive terms. The theories include appropriate rules of inference. All statements/theorems/equations derivable from the axioms are included in the Mathematical-Deductive System of a theory.

(II)   The concepts and results of the metatheory of mathematics apply to the set of physical theories.

(III)   To each physical theory there corresponds a set of physical phenomena. The theory's phenomenology is the set of all experimentally true statements/theorems/equations (STE's) in the physical theory. The nature of the inductive process makes it possible to disprove an STE but usually not to prove an STE. Therefore, we will define the *phenomenology* to be the set of all STE's in the theory minus the set of STE's that are experimentally proven to be false. We assume <u>all</u> possible experiments have been performed on the theory's phenomena so the set of false STE's is fully determined.

The close relationship of mathematics and physics make axioms I and II readily acceptable. Axiom III is physically reasonable. Some mathematicians have adopted a philosophic position that is close to the statement that mathematical constructs only have a reality if they correspond to physical phenomena or to alternative constructs of physical phenomena.

13.4 Definition of Primitives on Axioms

The primitives of physical interest in the known fundamental particle physics theories are quantum fields and space-time coordinates in the Standard Model type of theories, and Superstring fields and coordinates in Superstring theories. It is possible (probably likely) that a more fundamental form of Theory of Everything will eventually be developed. An issue that may then arise is the precise definition, and possible alternate definitions,[22] of the primitive terms. The development in this book does not require a detailed discussion of primitive terms so this issue will not be considered here. Section 15.6.2 does describe subsets of physical theories generated by changes in the definition of primitives.

13.5 Phenomenology – The System (Set) of Experimental Statements

As stated in axiom III there is a set of physical phenomena associated with each physical theory. We defined the set of all experimentally true statements (STE's) as the *phenomenology* of the theory. This set has an infinite number of elements for all but the simplest theories. The empty theory has an empty phenomenology.

The nature of scientific experiment and of the inductive process makes it possible to disprove an STE by finding one experimental result that does not satisfy the

---

[22] See p. 192 of Kleene (1967) for a discussion of an example of the effect of the definition of primitives on a mathematical-deductive system.

STE. The experimental "proof" of an STE is almost never possible since most STE's depend on variables that can take an infinite number of values and thus would necessitate an infinite number of experiments to fully prove them. As it turns out most mature correct theories have enough verified interwoven STE's in the range of their phenomena that their STE's can be taken to be true and correct. The only major exception to this statement is Superstring theories, which, at this point in time, do not have any positive or negative experimental evidence. We include these theories in the set of all physical theories since they have not been disproved.

More generally, we will define the phenomenology of a theory to be the set of all experimentally testable STE's whether they are true or false or of unknown truth minus the set of all STE that are experimentally proven to be false:

{phenomenology of a theory} =

　　　{set of all experimentally testable STE's of the theory} –

　　　　　– {set of all experimentally testable STE's of the phenomena of the theory that have been experimentally shown to be false}

We assume all physically possible experiments[23] for all possible physical situations have been performed on the phenomena of the theory in question. An example of a physically impossible experiment is an experiment that would require more energy than exists in the universe.

An STE is experimentally testable if a physically possible experiment, or set of experiments, can be performed to test its validity.

We conclude that for each theory in the set of mature physical theories there is a well-defined phenomenology by assumption.

The first issue we will address is the relation between the STE's of the mathematical-deductive system of a theory and the STE's of the phenomenology of the theory.

Lemma 13.5.1: The set of experimentally testable STE's in the mathematical-deductive system of a complete theory for a set of physical phenomena is the same as the set of STE's in the phenomenology of the theory.

Proof:
1. By III every STE in the mathematical-deductive system of a theory that is experimentally testable is in the phenomenology of the theory.
2. To prove that an STE in the phenomenology is in the mathematical-deductive system of the theory we assume the contrary – one non-tautological, experimentally testable

---

[23] This assumption applies both to <u>real</u> STE's and to <u>ideal</u> STE's as defined for statements earlier in this chapter.

STE exists in the phenomenology that does not exist in the mathematical-deductive system. Then the mathematical-deductive system does not describe all the physical phenomena that the theory is supposed to describe, and is thus incomplete – contrary to assumption.

3. Therefore the set of STE's of the mathematical-deductive system of a complete theory is the same as the set of STE's of the phenomenology of the theory.

Corollary 13.5.2: The set of experimentally testable STE's in the mathematical-deductive system of a complete theory for a set of physical phenomena is in one-to-one correspondence with the set of STE's in the phenomenology of the theory.

The corollary is easy to prove by assuming it is not true and showing a contradiction results.

13.6 The Relationship of the Three Parts of the Metatheory of Physics

The three systems associated with a physical theory are its mathematical-deductive system, its calculus and its phenomenology as illustrated in Fig. 13.6.1.

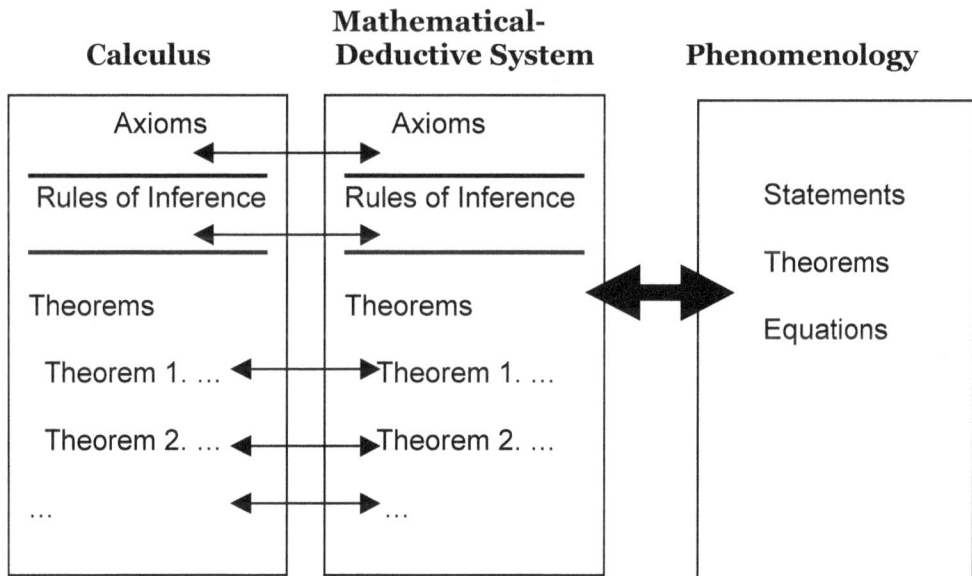

Figure 13.6.1. The three systems associated with a physical theory: its mathematical-deductive system, its calculus and its phenomenology.

From our preceding discussions it is clear that a "one-to-one-to-one" relation exists between corresponding STE's in the three systems. Thus we have developed a metatheory of physics theories in which each physics theory is composed of these three component systems. Based on this formulation, we will show that Gödel's Undecidability Theorem implies that the Theory of Everything T_E, if it exists, must be a quantum theory in the next chapter.

13.7 Consistency and Completeness of a Physical Theory

Consistency and completeness are major issues in any mathematical-deductive system. The consistency of a mathematical-deductive system in metamathematics is usually addressed by the consideration of its calculus (formal system). A calculus is *inconsistent* if it contains any formula F that is proved true through one chain of proof and also is proven false through another chain of proof. Consequently both F and Not F are true for some formula F in an inconsistent calculus (and thus the corresponding mathematical-deductive system):

$$F \text{ \& Not } F$$

If the calculus includes the propositional logic and is inconsistent, then all well-formed formulas that one can construct from the variables and operators of the physical theory can then be proved to be correct. A physical theory, which implies any syntactically correct formula is correct, is of course unacceptable.

The problem of proving inconsistency is similar in concept to proving an experimental prediction is wrong – one must find a formula F that satisfies both F and Not F are true – just as one must find the experiment that proves an experimental prediction is false.

In the case of physical theories the issue of consistency is usually a subtle one if present. Generally the axioms of a mature physical theory are consistent. However, there are cases where the axioms appear consistent but the theorems/predictions of the theory reveal an inconsistency. An example is Lee-Wick Quantum Electrodynamics which appeared to solve the problem of the divergences of conventional QED at the price of introducing negative metric states and negative probabilities that were then banished using a Dirac metric operator formalism. A study of higher order perturbation theory diagrams revealed the presence of negative metric intermediate states – an inconsistency in the theory since negative metric states were supposed to be eliminated by the Dirac formalism. The inconsistency was subtle. The theory was abandoned.

More generally, Gödel has also proven an "Unprovability of Consistency Theorem" that suggests that a consistency proof of complex physical theories such as a T_E may not be possible. A simplified statement of Gödel's "Unprovability of Consistency Theorem" is *"If a calculus C is consistent, then C's consistency cannot be proven within C."* Gödel's proof of this theorem is based on the use of constructive methods of proof (as opposed to non-constructive methods of proof such as transfinite induction that are not part of the constructive Rules of Inference.) The lack of a proof of consistency may not be a major issue for the T_E, if it exists, since inconsistencies tend to surface rather quickly in physics theories of the type with which we are familiar.

The question of *completeness* is a difficult one – especially in the case of quantum theories. A mathematical example of the difficulty of proving completeness is the fifth postulate of Euclid's geometry. Geometers wrestled for centuries with whether it was an independent axiom or provable from the other axioms. In the 19th century non-Euclidean geometries were invented which demonstrated that the fifth postulate was required to complete Euclidean geometry.

In most physical theories the question of completeness is usually settled by a study of the phenomena. If all possible experimentally testable predictions are made by the theory for its phenomena domain, then it is presumed complete.

In the case of a Theory of Everything T_E it is possible that a situation similar to Euclid's fifth postulate might develop. The independence of the set of axioms of the T_E might be in question. Perhaps one or more of the axioms might be derivable from the others. Judging from the case of Euclid's fifth it might then become necessary to study other candidate T_E's to resolve the issue of the independence and completeness of the axioms. Unfortunately, there is a practical issue: there are no other universes available (as far as we know) to use in phenomenologies for the other T_E's.

13.8 Is a Theory of Everything T_E Possible?

In mathematics the theorems of Gödel show that no one formal theory of mathematics exists.[24] Since a T_E is a mathematical-deductive system that in principle ultimately applies to all physical phenomena one can ask whether a T_E is actually possible. The parts of mathematics included within current candidates for a T_E include parts of mathematics that are in doubt within the framework of theories of the nature of mathematics and the nature of mathematical proof.

Under these circumstances if a generally acceptable, detailed, candidate T_E surfaces, profound philosophical, and metaphysical, questions will arise as to its nature, meaning and ultimate provability. Therefore it is unlikely that the "true" T_E will be rigorously provable to be a theory of everything until fundamental issues concerning the nature of mathematics are resolved.

Physicists, being what they are, will probably ultimately have a T_E that "works" like the Standard Model "works" for electroweak and strong interaction phenomena, and that situation will be physically satisfactory but not necessarily intellectually satisfactory. Some issues that may still be present are described in the following subsections.

---

[24] p. 14 Curry (1976).

13.8.1 Ambiguous Definition of Primitives?

The definition and nature of the primitive terms will raise physical, metatheoretic and philosophic questions that might be difficult to resolve.

13.8.2 Disconnected Theories of Everything

It is possible that the true T_E may consist of several parts that do not interact with each other at all. Thus we could have universes with different "sub-T_E's" and with no knowledge, or possible knowledge, of each other.

The T_E that we discover will then be only part of the answer.

13.8.3 "Many Universes" Theories of Everything

There are physics theories that posit the existence of many universes. Some many-universe quantum theories assume a new universe appears every time a quantum measurement is made. These theories, to be acceptable, would have to provide for interactions between these universes. Otherwise we would be unable to physically confirm any theory of this sort since the phenomenology would be incomplete.

13.8.4 Infinite Number of Levels of a Theory of Everything?

Some physicists have raised the possibility that physics consists of an endless chain of ever more fundamental layers. Currently we have the everyday physics layer, the molecular layer, the atomic layer, the nuclear layer, and the quark layer. Some have suggested that quarks, in turn, are composed of more fundamental particles. Some have suggested that the next layer is that of some form of superstring. An important point about this layered view of nature is that each layer becomes evident at a much higher energy than the preceding layer. Thus there must be a finite number of layers since the energy available to probe deeper is limited by the total energy of the universe, and, practically, is much, much less.

Therefore there is an ultimate layer, and at that layer the T_E, if it exists, must reside.

13.8.5 Philosophic Issues of Reality in the Metatheory of Mathematics

The nature of mathematics is a matter of some debate. Two of the dominant schools of thought are called the formalists and the platonists. The *formalists* view mathematics as more a matter of method than content. Formalists believe that mathematics deals with symbols but that there is no particular uniqueness to the choice of the set of symbols. One set of symbols is as good as another.

The *platonists* believe that the major concepts of numbers and sets have a reality in themselves apart from our understanding of them. It appears that most mathematicians and physicists favor this view.

Thus we see the question of what is reality cropping up in mathematics. Is mathematical reality method or substance?

13.8.6 What is Ultimate Reality?

Ultimately the same issues arising in mathematics will appear in the T_E when it is found. If the T_E takes the form of a quantum field theory then the particle fields would appear to constitute the fundamental reality. If the T_E takes the form of a Superstring theory, or a theory that involves geometrical constructs of a sort other than strings, then the issue that arises is: Is the fundamental reality mathematical form only? Or is the fundamental reality a type of substance (say strings) from which space, time, matter and energy are formed?

Since we cannot answer this question decisively in the case of mathematics it appears that a similar problem will appear for the T_E.

14. Gödel's Theorem Implies that Nature is Quantum Probabilistic Rather than Deterministic

neither v Gen r nor Neg(v Gen r)
K. Gödel

14.1 Einstein's Question

Einstein raised the interesting question, "Did God have a choice when he crafted the physical laws of the universe?" He pondered this question from time to time but did not announce an answer or even a partial answer.

The answer to this question requires more than just a Theory of Everything. It requires an analysis that includes all potential Theories of Everything – all the-might-have-been's for a Theory of Everything. Since the set of all possible Theories of Everything is certainly infinite, the task can only be partly performed at best by the development of general theorems that constrain the set of possible Theories of Everything.

One theorem that we will develop in this chapter asserts that the Theory of Everything T_E must be a quantum theory. This severely limits the set of possible Theories of Everything. Hopefully, other theorems will emerge that will further narrow the possible choices for a Theory of Everything. If Einstein's (and most particle theorist's) hopes are realized the range of choices could eventually be narrowed down to one theory – THE T_E. (It must be remembered that Einstein devoted most of the latter part of his life to a search for a unified theory that, in his view, would be the T_E.)

In this chapter we shall attempt to make a beginning on the program of finding constraints and requirements for the T_E. We will show that a restriction exists on the form of fundamental physical theories if we require that a fundamental physical theory predict the results of any possible experiment in principle. The key ingredient in the proof will be Gödel's celebrated Undecidability Theorem.

Before proving that the T_E must be quantum it will be necessary to determine the nature of the phenomenology of deterministic and quantum theories.

14.2 The Theory of Everything must be Quantum

There are several preliminary observations that will be needed to prove that the T_E must be quantum. The scenario for the discussion is based on the set of formulas that constitute the calculus. As stated earlier the set of calculus formulas is given by

{calculus} = {axioms} + {derived formulas} + {all other well-formed formulas}

= {decidable formulas that are true} + {undecidable formulas}

where { ... } indicates a set.

We now turn to the relation between a calculus and the experimental testability of the corresponding physical STE's. It appears reasonable to make the following assumption:

Assumption 14.2.1. All formulas in the calculus of a deterministic physics theory are experimentally testable either directly or indirectly in the sense that the corresponding STE's expressed in terms of the primitive and defined terms of the physics theory are experimentally testable either directly or indirectly.

The indirect experimental testing of an STE is the experimental testing of all consequences of the STE (derived from it and axioms and theorems). If none of the consequences of the STE are experimentally false then the STE is presumed true and included in the phenomenology of the theory. (We recall the phenomenology of a theory = {all well-formed, experimentally testable, statements not proven to be experimentally false after all possible experiments performed}.)

In the case of a quantum theory the relation between its calculus and the experimental testability of the corresponding physical STE's is based on the assumption:

Assumption 14.2.2. All formulas in the calculus of a quantum physics theory that only contain observable variables are experimentally testable either directly or indirectly in the sense that the corresponding STE's expressed in terms of the primitive and defined observable variables of the physics theory are experimentally testable either directly or indirectly.

The difference in the deterministic and quantum theory assumptions is due to the observability restrictions of quantum theories.

To prove that the Theory of Everything must be quantum we need one lemma.

Lemma 14.2.1 The calculus of each member of the set of mature physics theories meets the requirements of Gödel's Undecidability Theorem.

Proof:

1. Gödel's Undecidability Theorem is based on four assumptions: 1) Each recursive relation is definable in the calculus in the sense of proposition V of Gödel (1992); 2) The class of axioms and rules of inference are recursively definable; 3) The calculus is ω-consistent.[25] 4) The logic of the *Principia Mathematica* of Whitehead and Russell.

2. The calculus of each mature physics theory satisfies the above stated assumptions and thus Gödel's Undecidability Theorem.

∎

Theorem 14.2.1 The ultimate fundamental Theory of Everything (T_E) must be quantum.

Proof:

1. If we assume the T_E is deterministic then by Assumption 14.2.1 the set of experimentally testable STE's contains the set of theoretically undecidable statements since the STE's corresponding to the calculus of T_E includes the set of undecidable STE's.

2. Therefore a deterministic T_E is incomplete since some experimentally testable STE's cannot be proven theoretically – contradicting the implicit assumption of a deterministic Theory of Everything – a theory that describes all physical phenomena.

3. If we assume the T_E is quantum then by Assumption 14.2.2 the set of experimentally testable STE's is limited to STE's that contain only observable quantities and that do not contain unobservables.

4. Therefore if the set of undecidable STE's is contained in the set of STE's that contain at least one unobservable quantity, then the set of experimentally testable STE's about observables is contained in the set of provable STE's in the T_E. (i.e. in the mathematical-deductive system of the physical theory.)

5. Therefore a quantum T_E is possible if the condition stated in 4 is met.

6. If the T_E is a hybrid theory the deterministic sector encounters the problem exhibited in statements 1 and 2 above, and therefore is incomplete in the deterministic sector. Thus the entire hybrid theory is incomplete and not a T_E.

---

[25] A calculus is *ω-consistent* if it contains no class-sign c(n) such that both c(n) is provable within the calculus for every Gödel number n of the calculus and such that Neg[n Gen c(n)] is also provable within the calculus. In words Neg[n Gen c(n)] means the negation of the generalization of c(n) with respect to n. A *class sign* is a formula in the calculus that depends on exactly one free variable – n in this case.

7. Since physics theories are deterministic or quantum or hybrid the only possible T_E is quantum.

■

Thus we conclude that the fundamental theory of Nature must be quantum due to Gödel's Undecidability Theorem. This result is perhaps the first substantial requirement that has been demonstrated for the fundamental Theory of Everything. [26] Further constraints on the nature of the Theory of Everything will eventually be found through the application of the metatheory of physics. General issues that may be addressable within this framework are the rationale behind the group theoretic structure of the Theory of Everything, the nature of causality, and whether the theory is local.

---

[26] This proof simplifies and removes defects in the proofs given in Chapter 13 of Blaha (2005a), and in S. Blaha, "Mathematical Metaphysics: A New Theory of Physics Theories With a Proof that Gödel's Theorem Implies the "Theory of Everything" Must be Quantum" (preprint, May 19, 2005).

15. Gödel Numbers and Physics

There is no branch of mathematics, however abstract it is, that may not be applied someday to the phenomena of the real world.
N. Lobachevskii

15.1 Gödel Numbers, Turing Machines and Physics

Kurt Gödel developed a map from the formulas of a calculus to numbers (now called Gödel numbers) that he used to great effect in the proof of his Undecidability Theorem in his celebrated 1931 paper.

In the preceding chapter we showed that the Undecidability Theorem has implications for the Theory of Everything. In this chapter we define Gödel numbers for Turing machines, Quantum Turing machines, Quantum Computers, and physics theories. Then we develop a classification scheme based on Gödel numbers for the set of all Turing machines, the set of all Quantum Turing machines, the set of all lagrangians, and the set of all non-lagrangian physics theories. In addition we develop a mathematical criterion for simplicity for lagrangian, and non-lagrangian, physics theories that replaces the subjective, anthropomorphic (and somewhat vague) view of simplicity commonly used in commentary on fundamental particle physics theories.

The definition of Gödel numbers for physics theories is a necessary beginning point for the meta-analysis of the calculus of physics theories and the developments herein set the stage for further developments in the metatheory of physics theories.

15.2 Deterministic Turing Machines and Gödel Numbers

In chapters 3 and 4 we developed the concept of non-deterministic Quantum Turing Machines and presented examples showing how to define probabilistic grammar Production Rules. In this section we will begin by associating Gödel numbers with deterministic Turing machines.[27]

Before defining the Gödel number of a Turing machine we will consider a special case – the Turing machine generated by the production rules on page 16:

$$S \rightarrow AB \qquad \text{Rule I}$$

---

[27] We follow the conventions of Davis (1982) p. 56.

$$A \rightarrow y \qquad \text{Rule II}$$
$$A \rightarrow Ay \qquad \text{Rule III}$$
$$B \rightarrow x \qquad \text{Rule IV}$$
$$B \rightarrow Bx \qquad \text{Rule V}$$

We begin by numbering the symbols of a Turing machine with odd numbers ≥ 3 following Davis (1982). The symbols in this Turing machine can be numbered

$$
\begin{array}{cccc}
3 & 5 & 7 & 9 \\
\updownarrow & \updownarrow & \updownarrow & \updownarrow \\
A & B & x & y
\end{array}
$$

We will call the number assigned to each symbol its *token* using a well-known term in the theory of compiler-compilers such as *yacc* – the UNIX compiler-compiler. *The set of allowed token numbers consists of all odd integers ≥ 3.* (A compiler-compiler is a software program that generates a language interpreter from a grammatical specification of a language. For example one could build an interpreter that took C language code and transformed (interpreted) it into assembly language code.)

15.2.1 Gödel numbers of Expressions and Production Rules
A string expression such as

$$AyyBx$$

which can be generated by the Turing machine example under consideration, is assigned a Gödel number by combining the token numbers of each symbol in the string. The token values become exponents of a product of prime numbers greater than one. For example the preceding expression has the tokens in order: 3 9 9 5 7 and its Gödel number is

$$gn = 2^3 3^9 5^9 7^5 11^7$$

If we assign the token number 11 to the symbol $\rightarrow$ in rule III above then the production rule $A \rightarrow Ay$ has the Gödel number

$$gn(A \rightarrow Ay) = 2^3 3^{11} 5^3 7^9$$

Consequently we make the following definition:

0. Definition: If E is an expression (string) of symbols, $\beta_1\beta_2\beta_3\beta_4 \ldots \beta_n$ and $v_1, v_2, v_3, v_4, \ldots v_n$ is the sequence of tokens corresponding to these symbols, then the Gödel number of E is the integer

$$gn(E) = \prod_{m=1}^{n} Pr(m)^{v_m}$$

where $Pr(m)^{v_m}$ is the mth prime number raised to the power v_m with the 1st prime number taken to be 2. If E is an empty string (containing no symbols) then $gn(E) = 1$.

Due to the form of the definition of Gödel numbers, and the consequent uniqueness of Gödel numbers based on the theorem that any whole number has a unique decomposition as a product of prime numbers, the following definitions and theorems are evident:

1. Definition: If $g = gn(E)$, then the inverse relation is defined to be $E = gi(g)$.

2. Theorem: If E_1 and E_2 are expressions such that $gn(E_1) = gn(E_2)$, then $E_1 = E_2$.

3. Definition: If $E_1, E_2, E_3, \ldots , E_n$ are a finite sequence of expressions (strings), then the Gödel number of the sequence is

$$gn(E_1, E_2, E_3, \ldots , E_n) = \prod_{m=1}^{n} Pr(m)^{gn(E_m)}$$

where $Pr(m)^{gn(E_m)}$ is the mth prime number raised to the power $gn(E_m)$ with the 1st prime number taken to be 2.

4. Theorem: The Gödel number of an expression is never equal to the Gödel number of a sequence.

5. Theorem: If $E_1, E_2, E_3, \ldots ,E_n$ and $E'_1, E'_2, E'_3, \ldots ,E'_n$ are both finite sequences of expressions and

$$gn(E_1, E_2, E_3, \ldots ,E_n) = gn(E'_1, E'_2, E'_3, \ldots ,E'_m)$$

then $n = m$, and for $k = 1, 2, \ldots, n$

$$E_k = E'_k$$

15.2.2 Gödel Numbers of Turing Machines

6. Definition: It T is a Turing machine, and E_1, E_2, E_3, ... , E_n are the *sequences* of symbols of the n (different) production rules of the Turing machine, then the Gödel number of the Turing machine can be defined to be

$$gn(T) = \text{Min} \prod_{m=1}^{n} Pr(m)^{gn(E_m)}$$

where $Pr(m)^{gn(E_m)}$ is the mth prime number raised to the power $gn(E_m)$, and where Min indicates the minimum is taken over all possible permutations of the order of the production rules (sequences of symbols) and all possible permutations of the assignment of tokens to the symbols in the production rules. (We assume permutations of the symbols in a production rule are not allowed. Otherwise the minimum would also be taken with respect to permutation of the symbols in the production rules.)

Thus if there are n production rules that use m symbols, then there are n!m! Gödel numbers that correspond to the production rules (assuming the token numbers consist of the set of odd numbers ranging from $3 + k$ to $3 + k + 2(m - 1)$ where k is an even number greater than or equal to zero. (We set $k = 0$ for the sake of convenience.) A unique Gödel number for a Turing machine is obtained by taking the prescribed minimum.

7. Theorem: If the Gödel numbers of two Turing machines are the same then they generate isomorphic languages.

From this point one can develop the general theory of computability, and obtain an abstract formulation of Gödel's Undecidability Theorem.[28]

15.3 Quantum Turing Machines and Gödel Numbers

Non-deterministic quantum Turing machines were described in chapters 4 and 5. The definitions and development of the preceding section 15.2 can be used (with a few modifications shown below) in the quantum case also.

There are two types of Quantum Turing Machines: factorable (section 6.4) and entangled (section 6.5).

15.3.1 Factorable Quantum Turing Machines

A factorable Quantum Turing Machine associates a probability amplitude with each production rule of its Quantum Grammar (section 4.3).

---

[28] See Davis (1982) p. 121-2.

The mapping of factorable production rules to Gödel numbers is a generalization of the deterministic production rule mapping described in section 15.2. We must generalize the Gödel numbers of production rules to include probability amplitudes.

First we note that probabilities are real numbers and probability amplitudes are, in general, complex numbers. We have to generate an equivalent set of integers from the complex amplitudes that can be used as tokens in the construction of a Gödel number. All physics experiments measure probabilities approximately. Thus physically measured probabilities are accurate to a certain number of places (base 10), which we will call the *truncation length* (denoted tr). Therefore, when we specify the probability amplitude for a production rule, we can truncate the real and imaginary part of the probability amplitude to tr places. We can then find the lowest common denominator D for the set of production rule probability amplitudes and express the amplitude for each production rule in the form of

$$A_{rule} = (n_{r\text{-}rule} + i n_{i\text{-}rule})/D$$

where $n_{r\text{-}rule}$ and $n_{i\text{-}rule}$ (and D) are integers.

Based on these observation we see that the probability amplitude of each production rule introduces four additional integers – tokens – in the specification of the Gödel number of a quantum production rule: $n_{r\text{-}rule}$, $n_{i\text{-}rule}$, D, and tr. Therefore we define the Gödel number of a quantum production rule as follows:

8. Definition: If E is a quantum production rule consisting of a string of symbols, β_1, β_2, β_3, β_4, ... β_n and v_5, v_6, v_7, v_8, ... v_{n+4} is the sequence of tokens corresponding to these symbols, and if we define

$$v_1 = tr$$

$$v_2 = D$$

$$v_3 = n_{r\text{-}rule}$$

$$v_4 = n_{i\text{-}rule}$$

then the *quantum Gödel number* of E can be defined to be the integer

$$qgn(E) = \prod_{m=1}^{n+4} Pr(m)^{v_m}$$

where $\mathrm{Pr}(m)^{v_m}$ is the mth prime number raised to the power v_m with the 1st prime number taken to be 2. *The set of allowed token numbers for symbols consists of all odd integers ≥ 3.* This definition has the virtue of always using the first 4 tokens for the probability amplitude. Some definitions and theorems that immediately follow from this definition are:

9. Definition: It T is a Quantum Turing machine, and $\mathrm{qgn}(E_1)$, $\mathrm{qgn}(E_2)$, $\mathrm{qgn}(E_3)$, ... , $\mathrm{qgn}(E_n)$ are the quantum Gödel numbers of the n (distinct) quantum production rules of the Turing machine, then the *quantum Gödel number* of the Quantum Turing Machine is defined to be

$$\mathrm{qgn}(T) = \mathrm{Min} \prod_{m=1}^{n} \mathrm{Pr}(m)^{\mathrm{qgn}(E_m)}$$

where $\mathrm{Pr}(m)^{\mathrm{qgn}(E_m)}$ is the mth prime number raised to the power $\mathrm{qgn}(E_m)$, and where Min indicates the minimum is taken over all possible permutations of the order of the production rules (sequences of symbols) and all possible permutations of the assignment of tokens to the symbols in the production rules. Thus if there are n production rules that use m symbols, then there are n!m! Gödel numbers that correspond to the production rules (assuming the token numbers of the symbols consist of a set of odd numbers ranging from $3 + k$ to $3 + k + 2(m - 1)$ where k is an even number greater than or equal to zero. (We set $k = 0$ for the sake of convenience.) The numbers for $n_{\text{r-rule}}$, $n_{\text{i-rule}}$, D, and tr for each quantum production rule are defined by the physics of the theory represented by the Quantum Turing Machine. (We assume permutations of the symbols in a production rule are not allowed. Otherwise the minimum would also be taken with respect to permutation of the symbols in the production rules.)

Definition 9 leads to the following:

10. Definition: If $qg = \mathrm{qgn}(T)$, then the inverse relation is defined to be $T = \mathrm{qgi}(qg)$.

11. Theorem: If T_1 and T_2 are Quantum Turing Machines such that $\mathrm{qgn}(T_1) = \mathrm{qgn}(T_2)$, then $T_1 = T_2$.

12. Theorem: The quantum Gödel number of a quantum production rule is never equal to the quantum Gödel number of a Quantum Turing Machine.

13. Theorem: If the quantum Gödel numbers of two Quantum Turing Machines are the same then they generate isomorphic languages in which corresponding input state-output state transitions have the same relative probability.

15.3.2 Entangled Quantum Turing Machines

This case is better handled within the framework of the quantum computer formulation since we cannot associate a probability amplitude with each production rule as we did in the previous case. Benioff, Deutsch and others have shown a universal quantum Turing machine can simulate any physical process and a sufficiently complex quantum computer can simulate a physical theory. Therefore we conclude that quantum Turing machines can simulate a fundamental theory of nature producing probabilities for particle transitions from input states to output states.

We can associate a Gödel number with each statement in the program (See section 12.10.) of a quantum computer in the same way that we associated a Gödel number with each production rule in section 15.2:

14. Definition: If S is a statement in a quantum computer program consisting of a string of symbols, β_1, β_2, β_3, β_4, ... β_n and v_1, v_2, v_3, v_4, ... v_n is the sequence of tokens corresponding to these symbols, then the Gödel number of S can be defined to be the integer

$$gn(S) = \prod_{m=1}^{n} Pr(m)^{v_m}$$

where $Pr(m)^{v_m}$ is the mth prime number raised to the power v_m with the 1st prime number taken to be 2.

A quantum Gödel number can be defined by the sets of sequences of statements comprising the quantum computer program. In chapter 12 we saw how to translate Assembly language and C language programs into quantum computer operator expressions. We considered the case of a truly quantum computer program and, in particular, the quantum program example

$$h(\ ) = \alpha h_1(\ ) + \beta h_2(\ ) + \gamma h_3(\ )$$

where α, β, and γ are constants and where each term $h_i(\ )$ transform an input pure state (eigenstate of all the number operators N_m) into an output pure state. Each $h_i(\ )$ consists of a product of factors which we will call its statements. Therefore we will make the reasonable assumption:

15. Assumption: Quantum Computer programs have the form:

$$C_p = \sum_{i=1}^{k} h_i(\,)$$

where the k summands are products of factors S_{ij} that we will call statements

$$h_i(\,) = \prod_i S_{ij}$$

Statements can possibly include integrations over momentum variables.

Based on this assumption we will define the quantum Gödel number of a Quantum Turing machine program as

16. Definition: If TP is a Quantum Turing Machine computer program consisting of m symbols and k summands, and the Gödel numbers of the quantum computer program statements are $gn(S_{ij})$, then the *quantum Gödel number* of TP is

$$qgn(TP) = \text{Min} \prod_i \prod_j Pr(m)^{gn(S_{ij})}$$

where $Pr(m)^{gn(S_{ij})}$ is the mth prime number raised to the power $gn(S_{ij})$, and where Min indicates the minimum is taken over all possible permutations of the order of the k summands (but not of the statements within a summand) and all possible permutations of the assignment of tokens to the symbols in the program. The token numbers of the m symbols consist of a set of odd numbers ranging from $3 + p$ to $3 + p + 2(m - 1)$ where p is an even number greater than or equal to zero. (We set $p = 0$ for the sake of convenience.) (We assume permutations of the symbols within the program statements are not allowed. Otherwise the minimum would also be taken with respect to permutation of the symbols in the program statements.)

Definition 16 leads to the following:

17. Definition: If $qpg = qgn(TP)$, then the inverse relation is defined to be $TP = qgi(qpg)$.

18. Theorem: If TP_1 and TP_2 are quantum programs such that $qgn(TP_1) = qgn(TP_2)$, then $TP_1 = TP_2$.

19. Theorem: If the Gödel numbers of two quantum programs are the same then they generate isomorphic languages in which corresponding input state-output state transitions have the same relative probability.

15.4 Classification of the Sets of Turing machines, and Quantum Turing machines

In the preceding section we have developed a classification scheme based on Gödel numbers for the set of all Turing machines, the set of all quantum Turing machines, and the set of quantum computer programs. Each of these sets can be divided into subsets based on the Gödel numbers of the members of the sets by Theorems 7, 11, 13, 18 and 19.

15.5 Gödel Numbers and the Set of All Lagrangians

We can divide the set of all physical theories into two subsets: the subset of physical theories that can be defined by a lagrangian[29] (Lset) and the subset of physical theories that cannot be defined by a lagrangian (NLset).

We assume each physical theory in NLset is defined by a set of axioms or fundamental equations that generate its mathematical-deductive system of STE's.

We assume that all physical calculations in each lagrangian-based theory (Lset) can be performed using a path integral formulation in the case of quantum theories, and a standard lagrangian formulation in the case of a deterministic theory. This will enable us to restrict our considerations to c-number lagrangians.

We further assume that each lagrangian in Lset is defined by a finite string of symbols. Recursive functions are allowed to appear in Lset lagrangians.

In this section we will associate a Gödel number with each lagrangian. We will also show "proof schema" can relate equivalent lagrangians with differing Gödel numbers to each other. Physicists often redefine fields for physical reasons causing a new lagrangian to be generated from a prior lagrangian. A common example is a Higgs field that is often set equal to a constant ground state value plus a field. The lagrangian, after substitution, is different in form from the original lagrangian and has a different Gödel number. The Rules of Inference (which include substitutions) that are used in mathematical proofs can be used in the present case to transform lagrangians to different forms although their physical implications are the same. These considerations lead us to define a set of subsets of Lset which we denote $LPset_i$ for i = 1, 2, ... Each of these subsets contain lagrangians that can be obtained from each other through the use of variable (field) substitutions and other Rules of Inference. Thus all lagrangians in a

---

[29] Kenneth Wilson speculated in the early 1970's that the physical theories generated by classes of lagrangians, if solved, might turn out to show the same behavior near critical points and perhaps in other phenomena. Thus a consideration of the set of all lagrangian theories might lead to physically interesting results.

subset LPset$_k$ are physically equivalent to each other. The elements of these subsets have different Gödel numbers and a chain of equivalent intermediate lagrangians leads from one lagrangian to another lagrangian of the subset. The original lagrangian, the intermediate lagrangians and the target lagrangian all have different Gödel numbers but they all embody the same physics and are in the same subset LPset$_k$.

15.5.1 The Gödel Number of a Lagrangian

We can associate each lagrangian in the set of lagrangians with a Gödel number. The definition is similar in form to earlier definitions in this chapter.

A lagrangian is an integral of a lagrangian density. Every lagrangian contains a set of symbols representing fields, indices, and mathematical operators including integration symbols. To define the Gödel number of a lagrangian we first expand all repeated indices and other short hand notations in the lagrangian into literal symbols. So, for example, instead of an abstract indexed expression such as $A^\mu A_\mu$ where μ ranges from 0 through 3 we use the expanded form $A^0A_0 + A^1A_1 + A^2A_2 + A^3A_3$. Each symbol and mathematical number is assigned an odd token number greater than or equal to 3. For example,

Token:	3	5	7	9	11	13	15	17	19
	$\updownarrow$	$\updownarrow$	$\updownarrow$	$\updownarrow$	$\updownarrow$	$\updownarrow$	$\updownarrow$	$\updownarrow$	$\updownarrow$
Symbol:	A	B	x	y	$\int d^3x$	(	)	+	2

Based on the expansion of all repeated indices and shorthand notations we define the Gödel number of a Lagrangian as follows.

20. Definition: Suppose the Lagrangian L is an expression (string) of symbols, $\beta_1, \beta_2, \beta_3, \beta_4, \ldots \beta_n$ (including field operator symbols, mathematical symbols, integral signs, the number of space dimensions, the number of time dimensions, space-time indices, and any other quantities that appear directly or indirectly in the Lagrangian); and $v_1, v_2, v_3, v_4, \ldots v_n$ is the sequence of tokens corresponding to these symbols, then the Gödel number of a Lagrangian L is the integer

$$gn(L) = \text{Min} \prod_{m=1}^{n} Pr(m)^{v_m}$$

where $Pr(m)^{v_m}$ is the mth prime number raised to the power v_m with the 1st prime number taken to be 2; and where Min indicates the minimum is taken over all possible permutations of the order of symbols in the lagrangian that maintain its well-formed nature and all possible permutations of the assignment of tokens to the symbols in the lagrangian. The token numbers of the n symbols consist of a set of odd numbers ranging from $3 + p$ to $3 + p + 2(n - 1)$ where p is an even number greater than or equal to zero. (We set $p = 0$ for the sake of convenience.) *The set of token numbers consists of all odd integers ≥ 3.* A *well-formed lagrangian* has its symbols ordered in such a way that their order conforms to the syntax rules of the operators in the lagrangian. If L is an empty string (containing no symbols) then $gn(L) = 1$.

Since any number has a unique decomposition as a product of powers of prime numbers the following Corollaries hold.

21. Corollary: If L_1 and L_2 are expressions such that $gn(L_1) = gn(L_2)$, then $L_1 = L_2$ (the Lagrangians are the same Lagrangian.)

22. Corollary: If the Gödel numbers of two Lagrangians are the same then they generate the same physical theory.

15.5.2 Subsets of Physically Equivalent Lagrangians

If we take an arbitrary lagrangian and generate all possible lagrangians from it through substitutions, definitions of new variables (and field operators if the lagrangian is for a field theory), and other Rules of Inference that are used to transform lagrangians to a new form, then the set of lagrangians so constructed constitutes a set of physically equivalent theories. (Certain additional changes are made in the case of theories with path integral formulations.) As stated, we denote these sets of physically equivalent theories as $LPset_i$ for $i = 1, 2, ...$ These sets comprise the set of all lagrangians Lset. The use of physically equivalent lagrangians is exemplified by a Higgs particle lagrangian which is transformed to a different form to show the effects of its constant ground state value.

Those subsets of Lset, $LPset_i$, that contain classical (deterministic) physical theories such as theories in classical mechanics require no special treatment. Goldstein (1965) describes various forms of transformations: Legendre transformations and other transformations between equivalent physical theories.

In the case of subsets of Lset, $LPset_i$, that contain quantum theories changes of variables (fields) require an additional modification in the path integral formulation of the theories. The path integral integrand must have a Jacobian factor that reflects the change of variables in the lagrangian in order for the original lagrangian theory and the transformed lagrangian theory to have the same physical implications. For example if the path integral for a certain lagrangian in a certain subset of lagrangian theories is

$$Z(J) = N \int D\phi \exp\{ i\int d^4y \, [\mathscr{L}(\phi) + J(y)\phi(y)]\}$$

where N is a normalization factor. Then a change of field variable from ϕ to ψ

$$\phi = \phi(\psi)$$

in the lagrangian necessitates the introduction of a Jacobian factor $\mathcal{J}(\psi)$ in the path integral integrand (except for linear changes of variable)

$$Z(J) = N \int D\psi \, \mathcal{J}(\psi) \exp\{ i\!\int d^4y \, [\mathcal{L}(\psi) + J(y)\phi(\psi(y))]\}$$

See Huang (1965) or Weinberg (1998) for examples in the case of Fadeev-Popov gauge fixing.

Thus we have shown that it is possible to define subsets of theories for deterministic and quantum lagrangians where each lagrangian in the subset embodies the same physical theory.

We can define a Gödel number for a subset LPset$_k$ as:

23. Definition: The Gödel number of a subset LPset$_k$ is the minimum of the Gödel numbers of the members of the subset.

Due to the unique decomposition of a whole number into its prime factors we have the lemma:

24. Lemma: If the Gödel numbers of two subsets of Lpset are the same then they are the same subset.

In the next section we will define a criteria for selecting the "simplest" lagrangian in a given subset. This definition will be of some importance since potential Theories of Everything are judged by their simplicity as well as their physical implications.

15.5.3 Simplicity Criteria for Physically Equivalent Lagrangians

The criteria of simplicity and elegance, which are much talked about in discussions of the Theory of Everything, are subjective, ambiguous, and anthropomorphic. The assignment of Gödel numbers to lagrangians enables us to create a mathematical definition of simplicity and, to some extent, of elegance since these criteria are usually subjectively related. A lagrangian's Gödel number is a measure of the number of symbols and the number of times each symbol appears in the lagrangian. Therefore lagrangians in fewer dimensions, with fewer species of particles or fields, with lower spin entities, and with fewer terms will have a smaller Gödel number. Based on this observation we define the simplicity of a lagrangian as:

25. Definition: For any two lagrangians L_1 and L_2 if $gn(L_1) < gn(L_2)$ then L_1 is simpler than L_2.

which implies the following lemma:

26. Lemma: The Gödel number of the subset $LPset_{TE}$ that contains the Theory of Everything is the Gödel number of the simplest lagrangian in the subset.

15.6 Gödel Numbers of non-Lagrangian, Axiomatic Theories

A non-Lagrangian, axiomatic physics theory is a physics theory that has a finite set of axioms defined in terms of a finite set of primitives. Each axiom is assumed to consist of a finite number of symbols.[30]

The Gödel numbers of non-Lagrangian, axiomatic theories are defined in a similar way to the definition of the Gödel numbers of a Turing machine (section 15.2.2). The decomposition of the set of all non-Lagrangian, axiomatic physics theories NLset into subsets of physically equivalent theories is analogous to that of lagrangian theories described in section 15.5.2. We can also develop a similar definition of simplicity for this case as well.

15.6.1 The Gödel Number of a non-Lagrangian, Axiomatic Theory

27. Definition: It A is a non-Lagrangian, axiomatic theory, and a_1, a_2, a_3, ... , a_n are the *sequences* of symbols of the n (different) axioms of A, then the Gödel number of the theory can be defined to be

$$gn(A) = \text{Min} \prod_{m=1}^{n} Pr(m)^{gn(a_m)}$$

where $Pr(m)^{gn(a_m)}$ is the mth prime number raised to the power $gn(a_m)$, where $gn(a_m)$ is the Gödel number of the axiom a_m (calculated from definition 0 on page 144) and where Min indicates the minimum is taken over all possible permutations of the order of the axioms, all possible orderings of the symbols in the axioms that result in well-formed axioms, and all possible permutations of the assignment of tokens to the symbols in the production rules.

Thus if there are n axioms rules that use m symbols, then there at least n!m! Gödel numbers that correspond to the production rules (assuming the token numbers consist of the set of odd numbers ranging from $3 + k$ to $3 + k + 2(m - 1)$ where k is an

---

[30] I am indebted to John Schwartz for suggesting that I consider the case of non-Lagrangian, axiomatic theories.

even number greater than or equal to zero. (We set k = 0 for the sake of convenience.) A unique Gödel number for A is obtained by taking the prescribed minimum.

28. Theorem: If the Gödel numbers of two axiomatic theories are the same then they generate the same physical theory.

15.6.2 Subsets of NLset Containing Physically Equivalent Theories

Given a set of primitives and a set of axioms that define a physics theory, we can create a mathematical-deductive system that describes the physical theory in a manner similar to that used to define a mathematical-deductive system for a mathematics theory such as the theory of whole numbers.

Suppose we now take the primitive terms p_1, p_2, … of the theory A and define a new set of primitive terms p_i' with

$$p_k = p_k(p')$$

and substitute the new primitives in the axioms creating a new theory A'. The new theory is physically equivalent to the original theory A but its mathematical-deductive system will look different.

Now consider the set of physical theories generated by all possible redefinitions of the primitive terms as above. This set of physical theories will be a subset of the set of all non-Lagrangian theories NLset. All theories in this subset are physically equivalent.

A moment's consideration leads to the realization that NLset is composed of subsets of equivalent axiomatic theories which we will denote $NLset_i$ for i = 1, 2, … The elements of each subset have different Gödel numbers in general. Therefore we will define the Gödel number of a subset by:

29. Definition: The Gödel number of a subset $NLset_k$ is the minimum of the Gödel numbers of the members of the subset.

Due to the unique decomposition of a whole number into its prime factors we have the lemma:

30. Lemma: If the Gödel numbers of two NLset subsets are the same then they are the same subset.

In the next section we will define a criteria for selecting the "simplest" axiomatic theory in a given subset.

15.6.3 Simplicity Criteria for Subsets of NLset Containing Physically Equivalent Theories

A Gödel number is a measure of the number of symbols and the number of times each symbol appears in the set of axioms. Therefore sets of axioms for physics theories in fewer dimensions, with fewer species of particles or fields, with lower spin entities, and with fewer terms will have a smaller Gödel number. Based on this observation we define the simplicity of an axiomatic theory by:

31. Definition: For any two axiomatic theories A_1 and A_2 if $gn(A_1) < gn(A_2)$ then A_1 is simpler than A_2.

which implies the following lemma:

32. Lemma: The Gödel number of the subset $NLset_{TE}$ that contains the Theory of Everything, if it is axiomatic in nature, is the Gödel number of the simplest theory in the subset.

15.7. Escaping the Trap of Anthropomorphism

The form of Theories of Everything, and of physical theories in general, normally has anthropomorphic trappings: the choice of symbols, the form of the expressions, the characterization of the symmetries, and so on.

If we use the Gödel-number-based approach to defining lagrangian and non-lagrangian theories using the definitions of simplicity that we have developed we arrive at a non-anthropomorphic characterization of physical theories.

Thus we can hope that we can arrive at a universal characterization of the Theory of Everything that would pass this test: if we meet an intelligent alien scientist we could compare our theories of everything based on Gödel numbers! If the theories are the same their Gödel numbers should be equal. (We assume that they define tokens in the same way.)

16. Towards an Ultimate Theory of Everything

Our beginnings never know our ends.
T. S. Elliot

16.1 Where We Are

Today we have two types of fundamental theory: quantum field theories and superstring theories. Quantum field theories – particularly the Standard Model variants – are consistent with most of the known experimental data of elementary particle physics. The problems of Standard Model theories that currently preclude a simple, elegant unified Theory of Everything – divergences and unusual symmetry patterns – are being addressed. Blaha (2005) showed how to eliminate divergences from quantum field theories including quantum gravity making a unified theory possible. Studies of extended symmetries such as technicolor raise the possibility of a resolution of the unusual symmetries of current theory within the framework of a wider simpler symmetry. So there is some hope of a unified Theory of Everything, as we currently understand "Everything" to be, through the evolution of the Standard Model type of theory.

Superstring theories were born in the early 1970's in the Veneziano-Suzuki formula and the realization that string theories can provide a theoretical basis for it. In the past thirty years superstring theories of many varieties have appeared and much technical progress has been made on the theoretical front. But experimental support is currently lacking. And theoretical developments appear to have led us deeper into a quagmire of competing superstring theories. Experimental data, which superstring theorists feel is[31] "just around the corner" when new accelerators become operational may provide the guidance necessary to select a subset of acceptable superstring theories. At the moment it appears that superstring theory is somewhat adrift.

Both superstring theories and extensions of the Standard Model suggest that new phenomena await us at higher energies than currently available. Thus our understanding of what constitutes "Everything" may well change.

---

[31] A popular saying of the 1930's was "Prosperity is just around the corner." The corner arrived twenty years later around 1950.

16.2 The Set of Theories of Everything

If this the best of all possible worlds, what are the others like?
Candide (1759) - Voltaire

In discussions of the Theory of Everything or the Final Theory most commentators have exclusively addressed the issue of the *one and only* theory of everything. However, when we reach the stage of having a candidate Theory of Everything then the issue of alternate possible theories of everything must emerge. Einstein's question is precisely whether alternate theories of everthing are possible.

In this section we will consider the set of all possible theories of everything, and consider questions and issues that emerge in its consideration. Let us denote the set of all possible theories of everything as S_{TE}. Our primary goal is to determine criteria, if they exist, for selecting the Theory of Everything that governs our universe. It would seem that all theories in S_{TE} must be considered before one can understand why one particular theory was selected. It is of course possible that there is a subset of theories of everything such that any theory in the subset would have been an acceptable theory of everything in our universe. In that case one must face the issue of the mechanism that led to the selection of the actual theory of everything of our universe.

16.2.1 Requirements for a Theory of Everything

The primary current issue facing a theory of everything is "What is everything?" At this time we do not know what the particle spectrum of the theory of everything is. Most current candidate theories of everything suggest more particles, in fact many more particles. These particles remain to be found experimentally. It is also not known what the complete set of forces (interactions) is. Many theories postulate additional interactions as well as additional particles at higher energies.

Thus the primitives of the theory of everything are not known, and it is possible, that they may never be completely known since there are practical as well as theoretical barriers to the energies that we can probe experimentally. One theoretical energy barrier is the total energy of the universe (admittedly a somewhat ambiguous quantity). There might be new phenomena at energies in excess of the total energy of the universe.

Therefore the primary issue in defining the theory of everything is identifying what constitutes everything.

If we assume that we have at some (possibly future) time identified all phenomena of a theory of everything so that we can define the primitives of the theory then the following requirements are evident:

1.) The Theory of Everything must be consistent whether it is defined with a lagrangian formulation or an axiomatic, non-lagrangian formulation.

2.) All measurable quantities must be finite when calculated in the Theory of Everything.

3.) The Theory of Everything must be quantum.

16.2.2 Possible Additional Requirements

There are other *possible* requirements for a theory of everything. These requirements may or may not be genuine since their validity is a matter of opinion at this time.

1.) The Theory of Everything should be fundamentally geometric with all features derived from geometric considerations. The theory of General Relativity in which the force of gravity is reduced to geometry illustrates this type of theory.

2.) The Theory of Everything must be maximally symmetric with dynamically generated symmetry breaking.

3.) The Theory of Everything must support the development of life and higher life forms. (Anthropomorphism).

4.) The Theory of Everything is lagrangian-based. (Suggested by the form of the Standard Model)

5.) The Theory of Everything might emerge through a form of self-organization? Perhaps there are some ultimate "monads" (a fundamental layer) that obey dynamical laws that generate the layer of particles and interactions with which we are familiar.

6.) The Theory of Everything may be selected from the set of all theories of everything S_{TE} through a "principle of least theory" that is somewhat analogous to the familiar principle of least action. One simple example of a possible principle of this type is "the Theory of Everything is the theory with the lagrangian (or set of axioms) with the smallest Gödel number that incorporates all the primitives required by the physics and respects all symmetry and syntax requirements."

These *possible* additional requirements are only put forward for discussion at this time.

16.2.3 Why was One Theory of Everything Selected?

The answer to this question is at present unknown. If the answer were known then we would most likely know the correct theory of everything – at least in part. Possible answers are 3, 5, and 6 in section 16.2.2. Other possible answers are random chance due to unknown initial boundary conditions at the point of the Big Bang, other sources of random chance, and external selection.

16.3 Where We May Hope to Be

The analysis that we have gone through, particularly in the development of the metatheory of physics theories, suggests that we can move forward in certain areas in our understanding of the nature of an ultimate Theory of Everything. The theoretical development of the quantum computer paradigm for particle theories may deepen our understanding of the ultimate Theory of Everything.

Whatever the approach, it is clear that the ultimate Theory of Everything will raise profound questions as to the nature of reality, and of ultimate physics, that are similar to the questions that have already arisen in the metatheory of mathematics. Curry (1977) describes current views of the nature of mathematics (Physics *assumes* the nature of mathematics is understood.) with the statement, "there is no unanimity on the subject" and goes on to say, "There are two main types of opinion in regard to the nature of mathematics. We shall call these contensivism and formalism. According to contensivism, mathematics has a definite subject matter of content; the objects with which its statements deal ... exist in some sense, and those statements are true just in so far as they agree with the facts. On the other hand, from the point of view of formalism, mathematics is characterized more by its method than by its subject matter; its objects are unspecified or, if they are specified, are such that their exact nature is irrelevant".

Contensivism is composed of two principal groups: platonists who view mathematical objects such as numbers as having a real existence, and intuitionists who view mankind as having a primordial intuition which the human mind uses to construct mathematical objects and which exist therefore in the human mind.

Thus the fundamental nature of mathematics is reduced to metaphysics. Are mathematical objects real? Or are the objects merely "placeholders" in a pattern of relationships and methods?

When and if The Ultimate Theory of Everything is found Physics will have to confront similar issues. Are particles (if particles are part of the ultimate theory) real or are they a manifestation of the interrelationships in the laws of the ultimate theory. Or more succinctly, does reality consist of objects or relationships? This question is a question of metaphysics. Physicists will, of course, attempt to design an experiment to determine the answer. Is there such an experiment? If so, what will that experiment be?

References

Bailin, D. & Love, A., 1994, *Supersymmetric Gauge Field Theory and String Theory* (Institute of Physics Publishing, Philadelphia, PA, 1994).

Bjorken, J. D., Drell, S. D., 1965, *Relativistic Quantum Fields* (McGraw-Hill, New York, 1965).

Blaha, S., 2000, *Cosmos and Consciousness* (Authorhouse, Bloomington, Indiana, 2000).

Blaha, S., 2002, *Cosmos and Consciousness Second Edition* (Pingree-Hill Publishing, Auburn, NH, 2002).

Blaha, S., 2005a, *The Equivalence of Elementary Particle Theories and Computer Languages: Quantum Computers, Turing Machines, Standard Model, Superstring Theory, and a Proof that Gödel's Theorem Implies Nature Must Be Quantum* (Pingree-Hill Publishing, Auburn, NH, April, 2005).

Blaha, S., 2005b, *Quantum Theory of the Third Kind: A New Type of Divergence-free Quantum Field Theory Supporting a Unified Standard Model of Elementary Particles and Quantum Gravity based on a New Method in the Calculus of Variations* (Pingree-Hill Publishing, Auburn, NH, May, 2005).

Bogoliubov, N. N., & Shirkov, D. V., Volkoff, G. M. (tr), 1959, *Introduction to the Theory of Quantized Fields* (Wiley-Interscience, New York, 1959).

Braithwaite, R. B., 1960, *Scientific Explanation* (Harper, New York, 1960).

Carnap, R., 1995, *An Introduction to the Philosophy of Science* (Dover Publications, New York, 1995).

Carnap, R., 2002, *The Logical Syntax of Language* (Carus Publishing, Peru, IL, 2002).

Cottingham, W. N. and Greenwood, D. A., 1998, *An Introduction to the Standard Model of Particle Physics* (Cambridge University press, Cambridge, UK, 1998).

Curry, H., 1977, *Foundations of Mathematical Logic* (Dover Publications, New York, 1977).

Davis, M., 1982, *Computability and Unsolvability* (Dover Publications, New York, 1982).

Donoghue, J. F., Golowich, E. and Holstein, B. R., 1992, *Dynamics of the Standard Model* (Cambridge University Press, Cambridge, UK, 1992).

Duhem, P., 1954, *The Aim and Structure of Physical Theory* (Princeton University Press, Princeton, NJ, 1954).

Feynman, R. P. and Hibbs, A. R., 1965, *Quantum Mechanics and Path Integrals* (McGraw-Hill, New York, 1965).

Gödel, K., 1992, Tr. Meltzer, B., *On Formally Undecidable Propositions of Principia Mathematica and Related Systems* (Dover Publications, New York, 1992).

Goldstein, H., 1965, *Classical Mechanics* (Addison-Wesley, New York, 1965).

Huang, K., 1992, *Quarks, Leptons & Gauge Fields Second Edition* (World Scientific, River Edge, NJ, 1992).

Huang, K., 1998, *Quantum Field Theory* (John Wiley, New York, 1998).

Kaku, M., 1999, *Introduction to Superstrings and M-Theory Second Edition* (Springer-Verlag, New York, 1999).

Kaku, M., 1993, *Quantum Field Theory* (Oxford University Press, New York, 1993).

Kleene, S. C., 1967, *Mathematical Logic* (Dover Publications, New York, 1967).

Maly, K. and Hanson, A. R., 1978, *Fundamentals of the Computing Sciences* (Prentice-Hall, Inc., Englewood Cliffs, NJ, 1978)

Polchinski, J., 1998, *String Theory* (Cambridge University Press, New York, 1998).

Quigg, C., 1997, *Gauge Theories of the Strong, Weak and Electromagnetic Interactions* (Westwood Press, New York, 1997).

Quine, W. van O., 1959, *Methods of Logic* (Holt, Rinehart, & Winston, New York, 1959).

Quine, W. van O., 1962, *Mathematical Logic* (Harper & Row, New York, 1962).

Révész, G. E., 1983, *Introduction to Formal Languages* (Dover Publications, New York, 1983).

Smullyan, R. M., 1995, *First-Order Logic* (Dover Publications, New York, 1995).

Tarski, A., 1995, *Introduction to Logic* (Dover Publications, New York, 1995).

Weinberg, S., 1992, *Dreams of a Final Theory* (Pantheon Books, New York, 1992).

Weinberg, S., 1995, *The Quantum Theory of Fields* (Cambridge University Press, New York, 1995).

Weinberg, S., 1998, *The Quantum Theory of Fields II* (Cambridge University Press, New York, 1998).

INDEX

∞, 58
ϕ^3 theory, 20, 38
ω-consistent, 140

2-state observables, 58, 59, 67, 85

accumulator, 90
addresses, 119
alien, 56
alphabet, 11, 13, 30, 33, 34
alphabets, 11, 34
annihilation of particles, 8, 12, 13, 27, 34
annihilation operator, 65
anthropomorphic, 142, 156
anti-commutation, 50, 59, 61, 62, 66, 84, 113
anticommutation relations, 60, 65, 82, 83
Artificial Intelligence, 52
atoms, 2, 4
axiomaticists, 124
axioms, 124, 125, 126, 127, 128, 129, 130, 131, 134, 135, 139, 140, 150, 154, 155, 156, 159

Benioff, 47
Berkeley, Bishop, 1
Big Bang, iv
bit, 48, 49, 53, 54, 55, 58, 59, 62, 63, 64
bits, 1, 48, 49, 52, 54, 59, 60, 62, 64, 65, 67, 91, 93, 97, 99, 100, 112
Blaha, 161, 162
Bohr model, 77, 78
Bose operator, 67
boson, 31, 70, 85
boson field, 67
bosonic strings, 80, 81, 82, 85, 86

Brouwer, L. E. J., 124

C language, 56, 119
C++ language, 56
calculus, 3, 124, 125, 126, 127, 130, 133, 134, 139, 140, 142
central processing unit, 90
Chomsky, 11, 14, 15, 16
classical probabilistic computer, 109
classical probabilistic Turing machines, 3
color, 5, 6, 31, 32
color quantum numbers, 5
color triplets of quarks, 5, 6
colors, 5
commutation relations, 59, 61, 65, 66, 67, 80, 81, 84, 94, 113
compactification, 86
compiler-compiler, 143
completeness, 135
computer language, 10, 11, 13, 16, 33, 55, 71, 72, 73
computer memory, 35, 41, 75
computer tape, 59
conformal, 70, 81
Conformal Field Theories, 87
conformal invariance, 81
conservation of matter, 2
consistency, i, viii, 60, 85, 124, 126, 134
contensivism, 160
continuum limit, 50, 54, 58, 61
Continuum Quantum Computer, 50, 51, 52, 53, 54, 55, 58
CPU, 90, 91
Creation and destruction operators, 59
creation of particles, 8, 12, 13, 33, 34
creation operator, 65

About The Author

Dr. Blaha received his Ph.D. in Theoretical Physics from The Rockefeller University (NY). He has written many books on numerous topics including books on physics, consciousness and philosophy: *Cosmos and Consciousness*; mathematical history: *The Life Cycle of Civilizations*; and on Java and C++ programming. He has developed a new approach to quantum field theory described in his book *Quantum Theory of the Third Kind* that enables a finite unified quantum field theory of the Standard Model and Quantum Gravity to be constructed. Recently he completed a major study of Cosmology: *Quantum Big Bang Cosmology: Complex Space-time General Relativity, Quantum Coordinates, Dodecahedral Universe, Inflation, and New Spin 0, ½, 1 & 2 Tachyons & Imagyons*. He has served on the faculties of several major universities. He was an Associate of the Harvard University Physics Faculty for twenty years (1983-2003). He was also at Bell Laboratories, a member of management at the Boston Globe Newspaper, a Director at Wang Laboratories, and President of Blaha Software Inc and Janus Associates Inc. (NH). Among other achievements he was a co-discoverer of the "r potential" for heavy quark binding developing the first (and still the only demonstrable) non-abelian gauge theory with an "r" potential (1974); first suggested the existence of topological structures in superfluid He-3 (1975); first proposed Yang-Mills-like theories would appear in condensed matter phenomena with non-scalar order parameters (1976); first developed a grammar-based formalism for quantum computers and applied it to elementary particle theories showing that quantum field theories are equivalent to quantum computer languages(2000); first developed a new form of quantum field theory without divergences solving a major 60 year old problem and enabling a unified theory of the Standard Model and Quantum Gravity without divergences to be developed (2003); first developed complex General Relativity based on analytic continuation from real space-time (2004); first developed a generalized non-homogeneous Robertson-Walker metric that enabled a quantum theory of the Big Bang to be developed without singularities at t = 0 (2004); first generalized Cauchy's theorem and Gauss' theorem to complex curved multi-dimensional spaces(2004); first developed a physically acceptable theory of faster-than-light particles tachyons of any spin (2004); first showed a universe with three complex spatial dimensions has an icosahedral symmetry (2004); first developed the method of the composition of extrema in the Calculus of Variations; first suggested that inflationary periods in the history of the universe were not needed (2004); first proved that Godel's Theorem implied Nature must be quantum (2005), and first developed a quantitative model of the life cycle, and interactions, of civilizations (2002). He developed stochastic models for project economic analysis (1981) at Bell Labs. Blaha was also pioneer in the development of UNIX for financial and scientific applications, database benchmarking, and networking(1981-2). He benchmarked various Unix versions (1981) and found that Berkeley UNIX was twice as fast as Bell Labs UNIX which led to the conclusion that UNIX speed was bound by block size which, in turn, led to modern high speed UNIX with large block sizes. He pioneered Desktop Publishing(1985); and the use of C on PC's(1983). He developed a hybrid shell programming technique (1982) that was a precursor to the PERL language. He received Honorable Mention in the Gravity Research Foundation Essay Competition in 1978 for a paper describing a unified theory of the strong and gravitational interactions based on a non-abelian Kaluza-Klein theory with extra dimensions. He was nominated for three Awards for Technical Excellence in 1987 by PC Magazine for PC software products that he designed and developed. His email address is sblaha777@yahoo.com.

www.ingramcontent.com/pod-product-compliance
Lightning Source LLC
Chambersburg PA
CBHW081528220326
41598CB00036B/6367